The Year of No Summer

THE YEAR OF NO SUMMER

A RECKONING

Rachel Lebowitz

BIBLIOASIS
WINDSOR, ONTARIO

FIRST EDITION

Library and Archives Canada Cataloguing in Publication

Lebowitz, Rachel, 1975-, author

The year of no summer / Rachel Lebowitz.

Issued in print and electronic formats.
ISBN 978-1-77196-219-3 (softcover).--ISBN 978-1-77196-220-9 (ebook)

I. Title.

PS8623.E394Y43 2018 C814'.6 C2017-907000-2
C2017-907001-0

Edited by Stephanie Bolster
Copy-edited by Allana Amlin
Typeset and designed by Chris Andrechek

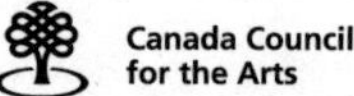

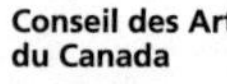

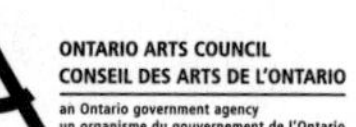

Published with the generous assistance of the Canada Council for the Arts, which last year invested $153 million to bring the arts to Canadians throughout the country, and the financial support of the Government of Canada. Biblioasis also acknowledges the support of the Ontario Arts Council (OAC), an agency of the Government of Ontario, which last year funded 1,709 individual artists and 1,078 organizations in 204 communities across Ontario, for a total of $52.1 million, and the contribution of the Government of Ontario through the Ontario Book Publishing Tax Credit and the Ontario Media Development Corporation.

PRINTED AND BOUND IN CANADA

For my son, Kaleb

They sat for a long time under the fig tree, listening to the familiar roaring of the sea. It was the same sound made by the forest back home. The child and the nurse thought about the world and how everything in it is related.

It is the kind of idea that comes later to most people. Decades past, one walks through a crowded room in which someone has died, and suddenly one recalls long forgotten words and the roar of the sea. It's as if those few words had captured the whole meaning of life, but afterwards one always talks about something else.

– *Embers* by Sándor Márai
(translated by Carol Brown Janeway)

"I once wept, sitting in that chair."
"For what reason?"
"I don't recall."
"Your memory fails?"
"Everything fails, sir. Reason, and harvests, and the human heart."

– *The Giant, O'Brien* by Hilary Mantel

Here's the weather, there's the war, here's the weathered dead.

Contents

Their Useless Wings

My God, my God, why hast thou forsaken me? why art thou so far from helping me, and from the words of my roaring? O my God, I cry in the day time, but thou hearest not; and in the night season, and am not silent.

We wait for thaw. For every tree is here deleafed or with dried cocoons hanging since November, so stubborn in their death. The branches brown with small black leaves. These sparrows are like sparrows, laced pattern of ice upon their wings. Housetop and ice and a heart that stopped.

O my God, dung like flakes of snow, a story here, a yelling and a remembering. A bringing in of birds, orioles from apple trees. Room full of wings and without, a darkness that will not let up. What are we to do in this night season? Lift our eyes. Open our mouths. Let dung melt on our tongues.

*

Frost on the window, lace on the wing. Etch designs on the glass with a thimble.

There is no God. There is no God and He is hurting us.

*

I want to talk about birds, about how many of them there were, of the passenger pigeons that went forth and multiplied into pies. *The banks of the Ohio were crowded with men and boys,* wrote Audubon, *incessantly shooting at the pilgrims, which… flew lower as they passed the river. Multitudes were thus destroyed. For a week or more, the population fed on no other flesh than that of Pigeons, and talked of nothing but Pigeons…* [Though] *they are killed in immense numbers… no apparent diminution ensues.*

They drove three hundred hogs into the forests, ready to be fattened on the bodies of birds. In this dark wood there are wolves and lynxes and bears and houses made of dung and mud and feather: you could give one to a man to show he is a coward, as if birds are cowed more easily than most. All you can say is that they're blue-grey and stupid and plentiful and when they're in a fright, they fly in *undulating and angular lines, mounted perpendicularly so as to resemble a vast column, and, when high,* [are] *seen wheeling and twisting within their continued lines.*

This, I suppose, should evoke for me starlings in their murmuration; at first, it did. But now I see a greyish scuttled sky or below, the deadened column of soldiers, men wheeling and twisting in their sodden lines.

*

Mud squelch and midden hands. And now we hear the wolves howl and see the lynx and bear and fox sneak off with feathers still hanging from their jaws. And now we walk among the dead and mangled and pile them in heaps. Open our holes: glimpse of rotten teeth. Now rats and hogs feed on the remainder. The sun does not shine. He hath covered it with His wing.

*

Long ago, before you or I or even our grandmothers were born, there was a city where birds died. They left their homes in the distant forests and dropped onto the dirt. It was cold, and the sky was rose or blue or speckled with bodies: people looked up—great whoosh of air. Then are became were. Then a girl picked up a body and held it in a kerchief but cold still seeped, icicle through cloth.

The girl's hands were freezing. Ice grew between her fingers like webs, frost stuck her eyelashes together. Her lips were red and chapped and the bird she held was red, but still it was as if colour had been leached away. The leaves on the trees were black. The girl opened her mouth, glint of icicle teeth. The crowd whispered and backed away, making the sign of the cross and stumbling over the others.

*

And so and so and so. It was the thirtieth day of the war. Martha, the last passenger pigeon, was found at 1 p.m., dead in her cage in the Cincinnati Zoological Garden. It was the same day over 30,000 British soldiers enlisted and just eight days after angels wavered in the air at Mons and showed the world which side God was on. Some said they appeared just as shimmers of light or a luminous cloud, others like winged warriors holding crossbows they'd used in the battle of Agincourt, and that later, the skins of the Germans were found to be pierced with arrows.

And so and so, the boys of Kentucky thus, with *torches of pine-knots* and poles and guns. And the noise the birds made, of a *hard gale at sea*, limbs giving way with the weight of them and crushing hundreds beneath. You couldn't hear the sound of the guns, he said, and only the sight of the shooters reloading showed him the guns were being fired. *No one dared venture within the line of devastation.*

So this vast plain of sound and the silence after, O how it hurt, and O how the wounded and O how the vast machine rolled out and over the bodies. By sunrise those who were able to fly had disappeared. Martha was packed in ice and sent to The Smithsonian, where she would be stuffed and displayed for over 80 years, until sometime after the cremation of my own grandmother. The waves crashed against the rigging of the ship. It was the same day that the British at Néry were surprised in the thick mist by German guns, and three of their four guns failed but they kept firing with the one. And 295 men were killed and 630 horses, bellies swell, breath pants, in France this, in Kentucky that, and the guns guns guns guns guns. Pack 'em in ice or scoop 'em into sandbags. A luminous cloud shines but on what? It was still a muddy war.

*

There are wilde Pigeons in winter beyond number or imagination, my self have seene three or four hours together flockes in the aire, so thicke that even they have shadowed the skie from us.

Explosions on April 5th, 1815: British soldiers in the Dutch East Indies think the military posts are under fire and they set out to search for pirates. On the 10th, Tambora, just twenty-six natives left of 12,000. The screaming the limbs the timber the sea the lava the houses the harvest the silence—ashes ashes, we all: dark.

Pumice a foot thick and ships blocked by boulders years later. Sea fills houses, houses become sea. Cows, people swept into sky and tossed back down again, heart fluttering in His palm: just a god's daughter's getting married, folks, these the explosions in her honour. Or: a sign, islands to be freed from your stinking rule, at last.

The lava, the horses, the runs, it's disease, distention, darkness for days. *Nothing equal to it was ever witnessed in the darkest night.*

Gods drink from calderas. Birds' wings are thunder. Chin Chin.

*

Pumice with fragments of skull wash up on beaches years later. Or perhaps a whole head, as flies are frozen in amber.

Wherever we go, there are people who tell us about God, or gods, what He means by this. What this broken body tells us, beyond its brokenness.

I question wisdom. I question all of it.

We bury the headstone. We play football with it.

*

The young men queuing up to enlist in 1914 have the look of ghosts. They are queuing up to be slaughtered: they are already dead.

When war was declared, the suffragettes stopped chaining themselves to fences and toured the country instead, giving speeches and white feathers to any man they saw in civilian dress. Man? I meant to say boy. Here's a boy who'd been in the retreat from Mons.

(Wherever God was, it was not there.)

Who fought in the Marne and in the first battle of Ypres, who caught a fever, who was sent home. Who was seen by four girls, who was given a feather, who *explained* [he'd]... *been discharged and... was still only sixteen. Several people had collected around the girls and there was giggling.* Who walked right into the nearest office and rejoined.

A young father got one as he was coming home from work. *That night he cried his heart out.* A pacifist said he'd gotten so many he had enough to make a fan. You could get enough to make wings, you could get enough to stick onto your outstretched arms, dying for some giggling girls' sins. The glue: hot tar or Passchendaele mud, slime of your best friend's guts.

And no birds sang? That's nonsense. Birds sang sang sang in music halls, pamphlets pushed against your chest.

*

We return to the birds falling on the streets of New York, falling at our feet as dead. Snow falling red *as the blood of a dead man.* And the birds falling dead in the field. And the nightingale and the lark. And the hummingbird, bluebird, robin, jay. Blackbird benighted, benumbed.

Ice in July, thick as window glass. *Great frost—we must learn to be humble.*

Passenger pigeons heading north suddenly veer south and we pickle them in hard cider. Caterpillars grow fur so long they can't crawl. Wolves descend. Bounties are proclaimed, as marriage banns; we leave church with hat and mitt and gun.

Days pass. We count the weeks with blackened kernels of corn. Throw rocks over our shoulders. Feathers fall onto red snow.

Yet is this covenant kept. *By the breath of God frost is given, and the breadth of the waters is straitened.*

I have a straightjacket made of the frozen sinews of birds. I have a scold's bridle made of the bones of a crow. The birds drop dead, and it is not our doing. It is not. We must learn to be humble. We must learn to be humble. We must—August 31st of this Year of Our Lord: mile-long iceberg near the Grand Banks, Newfoundland. Seagull squall overhead; below, our mittened hands clenched in prayer.

*

But see! The birds alight on smoking chimneys. Shimmer of orange and grey.

There's a man who nabs numb orioles and brings them to warm at his hearth. His hand cups this still body—this still-fierce beating heart.

Mud

The standing stones look like headstones. Soldiers' graves, maybe, from the Great War, the way they stand stark in the dusk. The setting sun glows behind the tallest one. All you see is a shimmer of light. The clouds are pulled apart.

I took this photograph. I stood in the waning light in Orkney, while my husband mused nearby with our baby boy upon his chest. I know exactly where this picture was taken, I know the circumstances, the way the wind blew and how our son cried fretfully when it hit his face. And how resentful I felt, coming back here but no longer single, no goofy dancing around the stones with the other Rachel I'd met at the hostel, no drinking tea from Cesco's Thermos, no sitting with my own thoughts while our cabbie waited. I had a child now who needed me and it was bittersweet.

I walked away. I took more photographs. I wanted to get just the right one for our living room wall. And now I look at it in this room and I think, but. But it looks like the war, how did I not see that I was photographing the war? Something about the darkness, the stones rising up from a ground so black that you don't even see where the stones begin. The low hill behind. If you go closer to it, you'll see thousands of crosses, row on row.

As crowds of people on a subway car echo the next war. Hey folks. The catacombs in Paris: performing tonight as Auschwitz.

*

You hear about the mud. Mud, yeah, lots of mud. And the word is not so threatening is it? Mud pies, mud baths, I remember once I took a group of children to Bowen Island, and we swam in the water and found heaps of mud there, and all of us, even those self-conscious pre-teen girls, piled it on our faces and threw it at each other—the giggling, the screaming and the "oh shit, I mean shoot, we're going to miss the ferry, hurry up!" That's how much fun the mud was. Or was it clay? Is there a difference?

How trite this is, how I babble on, to get away from what I was going to say. To return. We are always returning, to this moment, to this war that no one now remembers. No one from the war is not not-alive now, just as no one was not not-alive then. That doesn't make sense. For one thing, there were lots left alive in the fields. And for another, why not just say "no one was alive"? But I'm trying to get at the no of it. *Never, never, never, never, never!* Lear cries, holding the dead Cordelia to his chest:

No, no, no life! Why should a dog, a horse, a rat, have life, And thou no breath at all? Thoul't come no more, Never, never, never, never, never!

No man's land. No no no life. And this Passchendaele mud. Just:

Just:

Just:

space after the colon:

*

But not clean like this, not white. Just black, grey, sepia, black, black. It envelops. Do you know what it's like, when you're about to nap and there's this light that's right behind your eyelids and it's this beautiful white? This is not that.

The photograph:

Six stumps to our left rise out of the mud, or rather are stuck there. You can't say "rise" because that makes it seem like Excalibur, like there's some sort of resurrection, and there's nothing hopeful here. You can't say "grown," because not only are things dead here, they have never been not not-alive. So let's say stuck. Stuck there in the mud, like this whole dismal world is some kind of mud pie of God's. Let's say He stuck those stumps in the mud like sticks, because the sticks were just lying there, and that's what you do when you're bored and you see sticks, or you make up some kind of game with them. Maybe a battle. Maybe some of those sticks were guns. Some dead sticks that made other dead sticks deader.

On sunnier days, He could have looked for ants and then fried them with His magnifying glass. Though we could have told Him: it's a waste of time. There are no ants here. Or sun.

When I first looked at this photograph, I thought: those six stumps are like dead limbs, like men's arms rising (stuck) in the mud. But that's so histrionic. History need not be histrionic. And I lie too about the living, because in the mud which I haven't even begun to describe and can't, there is a man, standing (stuck), a man who was once alive though he's long dead now. He is facing the camera. All around him are shell holes filled to the brim with rain.

It is Passchendaele, the word that evokes passion. Not this dreariness. *No, no, no life!* And no God either. If He were playing with sticks here once, He left his play, as He ran in for His lunch. And His mother reprimanded Him for His dirty hands. *Howl, howl, howl, howl!* O you are a god of stone.

Never, never, never, never, never! The screen door slams as He rushes in—"hurry, hurry! We're going to miss it!" When we got on the ferry, I bought the kids ice-cream cups, those ones that come with the wooden paddle spoons. The mud under my nails came off when I did the dishes that night. The photograph is mute.

Second Battle of Passchendaele—16th Canadian Machine Gun Company.jpeg
Second Battle of Passchendaele—Field of Mud.jpeg
Second Battle of Passchendaele—Barbed Wire and Mud.jpeg
Second Battle of Passchendaele—Movement and Traffic.jpeg
Second Battle of Passchendaele—Wounded.jpeg

*

Barbed wire and mud, more shell holes. Gnarled wire, one strand sweeping up from what looks like nothing more than a bird's nest, to form a backwards "C." Grey horizon, land seeping into sky. Then this line in the air. Blue ballpoint. Who looked at this? Whose hand held the pen? Which library was this in? Soft swish as the photographs were turned. Hands encased in yellowed archive gloves. A pen that took notes? A pen that slipped. A pen that made a vertical line in the air, slight curve at its tail. Blue accident overlaying grey intent.

*

Nobody would believe that there could still be human beings in this churned up wilderness.

This once marsh. Then ducks, then reeds, then frogs, then bird. Heron. Then log, then sitting, then book, then look, finger marking your place: frogs, frogs, dusk. The mosquitoes are out.

Then claimed, drained, renamed. Then carrots, then potatoes, then snow peas, aubergine. Then mushrooms, then onions, fried in butter. Word, harvest. Word, home.

This farmland is our flag, it's how we do things. Call it pissing. Call it love. Because this mouth. Because this child. Because we can. Because we were given dominion. Because carrots, garlic, butter, potatoes. Because plums, plum-juice down her face. Because lips, because kiss. Because God told us. Because we're human. Because everything must eat.

*

But the marsh is not dead. Everything comes alive again. Bodies die three times over. This cemetery gets bombed out and bodies from years past are unburied. Tambora erupts and sunken prows surge onto the beach.

History is like scurvy: you get sliced, your wound heals, you break your leg, your tissue patches you up. But then rotten beef, rotten pork, mouldy bread, maggots: your gums turn black, they grow over your teeth like moss, grass, marshland. And the collagen, which binds things together, fails. So that arm you broke, that day when you played football, when you were 10? It breaks again. The man who bears from years before a battle scar, huddles in his hammock and watches the blood seep out, again. Your shattered legs shatter.

A farmer ploughs up a skull. A woodsman's axe clangs on a shell. And some long-dead cousin of mine never left the war. Children outside his window hoot, "Bang Bang You're Dead, Fifty Bullets in Your Head": and he's shivering, he's in that graveyard now. That was another war, that was another war, that "was"? That "was" is a lie.

The mud was six inches deep everywhere, and in most places halfway up to my knees. The surrounding country was... shot to pieces, looking like a field after trees and stumps have been pulled out, except that the holes [were] *as deep as 10 feet and filled with water. The lips of one shell hole practically* [touched] *the lips of another.*

Red lips are not so red. The drainage canals have been bombed out, the drainage cannot drain. This land is flooded mud.

Say what you will of intent. Yes, this was intended, this slick slump of mud, the thousands dead.

But we were fools in other ways. We thought: it's claimed, drained, renamed. We thought we changed it completely, we thought it's ours for keeps. *Veni, Vidi, Vici.* As if that's all we needed. As if that meant an end to it.

*

One Allied soldier killed
for every centimetre gained.

And still it rained.

*

(Think of Tambora's trees and the way He wrested out of hands, hands. Think of the next year: the rain falling for weeks. The Rhine rising to fifteen feet, seven inches. The villages that drowned. The hailstones that pelted crops; the hailstones that shattered windows.

It was a year of prayer and absolution. In Flanders, a thunderstorm. The cavalry sounded at 9 p.m.: blast of trumpets—cries, garments, lamentation. Hundreds rushed forth from their houses and fell to their knees.

Meadows were soaked. Farmers rowed out to cut the grass for fodder.)

Fodder, fodder, fodder, fodder.
"*Mutter mutter mutter.*"

*

Glass falling slightly. Weather unsettled. Heavy showers of rain throughout the day.

Abdominals abdominals abdominals abdominals.

Can God be on our side? Everyone is asking.

Our men (*brought in with mud over their eyes and mouths*)
Our men brought in with
Our men brought in
Our men brought
Our men
Our

And no one in the War Cabinet
to lift a voice in protest.

*

(*By the time you get this, it will be history.*)

*

But the thrushes got used to the guns. They sang above the mud and sticks of trees, above the hiss and scream of shells. They spread their wings over the dark waters. Cocked their heads just so.

On walls of slime, there were mirrors and pin-ups and clocks that ticked. On tables, playing cards lay stuck together in the damp. The walls were rivulets, bits coming off in the rain.

Best to speak, instead, of trees.

They were mostly dead, of course, black and gaunt, with rain dripping from their lifeless stumps and the mud sludging up. But here and there a leaf grew, which had nothing to do with hope.

You can see the craters now, lovely ponds and perfectly round, with grass growing up alongside and ducks gliding, and a clear reflection of trees and clouds. It's a beautiful day. The locals made quite a sale of items found near here: buttons, badges, pistols, boots. Holsters and wooden pipes. Fragments of shell. It cost one penny for a brass button. Twenty francs for a Smith and Wesson revolver.

Tambora

I dreamt I was outside. It was 5 p.m. and so black that I could see nothing. No streetlights, no sidewalk, no self. Not even a shimmer in the air where my body was, legs moving, arms swinging slightly at my sides. It's becoming winter, I thought. It gets so dark in November.

The dark was a complete thought. It was a solid thing. And yet it was characterized by the absence of light. Like atheism, it could only be defined by what it was not.

It was a clear night, but there were no stars. No moon. No God.

Though I say now the dark was solid, though I recall it as thick, substantial—a blanket—I felt no resistance. My body was not pushing against something. Though I was fumbling with my feet, with trying to figure out where my body was in relation to the sidewalk, the earth, with trying not to fall, the air itself was as light as it had always been.

The blackness was the black of black crayon. You take these colours, you take all the purples and pinks and yellows and blues and greens and you scribble them on the page. And then you take the blackest crayon and you cover the colours so you can't see them, you draw this pall, this *drawing-down of blinds*. And then with your fingernail, you scratch away some of it so you can see the colours underneath. This, when we were young, we called a magic slate. But if you chose not to scratch it off, it would have been like this.

I dreamt once that there were trains, and they were moving through the darkness. And you could hear the whistle through the dark. And the snow was falling. And something about leopards, their lithe bodies, catching up to the train. Were they like the wolves of fairy tales? Were we throwing out the bodies of our loves, to save ourselves? I cannot recall. Just the dark and the snow and the way the trains wound round the hills, serpentine, and the great giant leopards with their great giant padded paws. Then my son woke me. He was standing by my bed. His only words: "Why is the earth always changing?"

I think of that moment and this darkness and the way our bodies walk on the earth's crust or sit upon it in trains. And how dark it was then, and how dark it became, after Tambora erupted, and you couldn't even see your hand. At that moment, in the day, it was dark, darker than the darkest night.

I once worked with someone who believed, fervently, in God. On his last day of work there, he tried to convert me. "Imagine," he said, "imagine you are in a car factory, and all the lights are out. You cannot see a thing. And someone says, look, you are in a car factory, you are surrounded by parts of cars. But you cannot see them. But does that mean they aren't there? So it is with God."

The word Tambora is like a drum, "tambour." It rumbled. There was something beating beneath that mountain, something we think of as dark, because we don't see it. But is it? If there's no light that reaches the depths of a volcano, is the magma still orange? Is it orange covered by black crayon?

Was the darkness in my dream like blindness, or is blindness another colour, a slate grey, a white, a beige? It was so black. And still no stars, no moon, no God to light my way.

*

A "tambour" is also a circular embroidery frame, made of two interlocking hoops. It's a circular wall, supporting a dome or colonnade.

It's the rolling top of narrow strips of wood glued to canvas, like the roll of a rolltop desk. It's a shallow cup or drum attached to a thin elastic membrane and writing lever, used to register slight motions: the movement of your arteries or the contractions of your digestive tract.

It rolls up and down. It records the secrets of your body, the steady beat of your pulse, the food pushed through, through and down.

It's telling us that there is art and there is blood.

It's telling us that there are walls that are not straight and that this, too, is art.

That there's movement, and that there's stillness. That there's the roll of the rolltop desk and then the desk is still. That detritus is hidden, not gone.

That there is quiet. That there's this needle, these bright threads.

That though a woman is quiet, she's telling a story. This frame is holding the story.

That the woman holding the frame that is holding the story is also a body: that this body, at the same time, is holding a story. And that this tambour is recording it.

That what looks steadfast is not, always, and that what rolls down will stop.

That all of this is made. That drum, that frame, that wall, that rolltop, that medical device: all of these are ours. That the analogy to Tambora can only go so far.

There is a frame and then there's this.

*

Here's what happens. The top of the mountain blows off. When the magma chamber beneath partly empties, it can't support the weight of what's left. If Atlas truly left, at last. If the tambour slid from its domed roof. If I walked away. The part left standing can't.

Mommy, why is the earth always changing? Magma turns to lava. The mountain is a hole, no, not bottomless: a bowl, *caldària*, a cooking pot. Steam rises, mud bubbles. *Fillet of a fenny snake, In the cauldron boil and bake.*

This is one of the world's deepest calderas: four miles wide and 4,000 feet deep. At its bottom, stubborn life: trees, reeds, bugs. There's mud, but it's not too ominous. You can move through it fast enough on this bright day. What looks like patches of dried solid clay, caked into cracks and fissures, changes to liquid fast enough: you step on this jigsaw puzzle and it moves again into mud, gives easily, sticks to your pants.

There is a video and the people in it are tiny silhouettes, half the size of my pinkie. They walk across the earth. They pitch their tent. They move towards the lake. There is a lake. There's a stream, too, and the colour of the land is grey and black and brown and green. It's a muted beauty: so grand, so craggy, so unsubtle, but covered in this mostly black and grey palette so that the green is very green.

There's this sound of piano: crescendo, calando! because we can't watch without having our mades overlapping a stream lapping.

A cooking pot. Cast iron. You stir so the food doesn't burn. Such a tidy domestic metaphor for this unfathomable depth. I understand cooking pots. I understand soup and stew, and the importance of spoons. I understand trees, or at least, I understand what a gardener said: "They want to live."

I understand that really there is no contemplation about it. I understand that the tree is not ponderous, nor the reeds. I understand that the mud is not broken up because it's broken. I understand that the little pebbles in that stream are not nicer and sweeter than the boulders beside them, that the stream is not moving quickly because it's anxious to get somewhere. I understand that the steam rising from below is not cirrus clouds. It's not that earth and sky mated and He covered Earth with His body every night. It's not that this steam shows He's moving back up now, having peopled the world with giants and their earthenware: cups, bowls, plates, pots.

*

On April 10th, 1815, three columns of lava met in the air in a *troubled, confused manner.* Pumice fell from the sky: twice the size of an orange, others the size of walnuts.

When my son was inside me, he was likened, often, to food. Your baby is the size of a poppyseed. Your baby is the size of a blueberry, date, lime, plum. Peach, orange, mango, grapefruit. Till: pumpkin, watermelon. I think it was meant not just as a frame of reference that we could understand, but as a comfort—food, after all, keeps us alive. Food belongs in our bodies. Picture this instead: marble, bar of soap, brick, cow pie.

No one ever says your baby is the size of a bullet. Your baby is now growing to the size of a grenade.

If the child is likened to a plum, it is to a plum you have not yet eaten. It lies there on the dish, or hangs upon the tree, perfect, circular, unmaimed, unbruised, unbitten, unchewed. It's like the first snowfall of the season, covering the hills and dips of the earth. Your baby is a dollop of snow on the top of a branch. Your baby is this wide expanse before footprints mar it and the cars turn it black.

Food belongs in our bodies, but not like this. If we picture the baby inside us, it is as something that we have not touched or have swallowed whole, like Cronus did his young.

We like to think of things remaining whole, even when they've been eaten. Not this disarray, this pumice falling, this havoc, this cacophony.

Despite it all, I picture it as quiet. Even the ash raining down is silent, peaceful. Ash that's forty inches thick covers villages. There are dead fish on the surface of ponds, and this too is quiet, a considerate death: bodies floating, children on air mattresses.

Small birds fall dead from the skies. Their bodies make no sound on the pillows of ash. They remain whole.

It is so dark that you have to eat your lunch by candlelight, which is a lovely word. Trees are uprooted, true—

but I've gone from thinking of this to thinking of places where there are no trees to fall. I'm thinking of Orkney, again, and of the wind and stone and light, where the land was mostly flat and the few trees I saw had a dozen rooks' nests in each one. How it was so far from everything, how the grey light glimmered on the dark sea.

How nothing is far from anything.

How if I were to write about Tambora, properly, it would just be screams. Only how do you write screams? How do you write what it's like, to have a baby move inside you? It felt like nothing but itself. And we will say the same for—eruption! This suddenness, this breaking off, this tossing up, this mastication. It's all quick: no "*mutter, mutter, mutter*" on the battlefield.

Nearly all of the 12,000 residents of Tambora died within the first twenty-four hours. They died of ash and pyroclastic flows. They died from streams of fire. Death was dramatic and quick.

But beyond this, there were the others, who kept dying, who wouldn't stop dying, who wouldn't be dead. Low moan, pulsating brain, white bone, guts seeping out. They died slowly, of respiratory infections from the ash, of ash in the water, of no food but ash. 90,000 Indonesians starved.

I think how I was, often, just a body carrying a body, his body that looked nothing like a plum, but oh, the comfort of that, the perfection of it. And snow, the way it first appears. The way ash looks when it's cooled and grey, heaped beneath your fireplace grate.

*

I read there were no survivors in the kingdom of Tambora, or in the neighbouring kingdom of Pekat. I read too that there were twenty-six, and I wondered whether they were children or grandfathers or tired mothers, whether they grew rice or mung beans or both, whether they were eating dinner or were in the fields or playing clapping games when the first rumblings began. What happened to them? Did they die in the months after? Certainly they weren't made famous, nowhere are they paraded about or stuck in a living museum exhibit, carted about and poked, like Saartjie Baartman, who was to die of smallpox or pneumonia or syphilis just six months later. Where are their bones? Why are they not on display, like Baartman's? She was ogled for 150 years.

It's true that they weren't slaves, but neither were they white or free. How quaint with his native clothes and sombre face: fan yourself and sigh at those haunted eyes. The end of a kingdom and of a language, locks of hair pressed under glass. That the natives did not relish their situation is made clear by their gods, who blew up this mountain to signal the end to colonial rule. That there is no skin or fibula in museums, that these twenty-six were not captured even in print or sketch, with a caption underneath ("survivor of the worst volcanic eruption in recorded history"), that we have records of the size of the lava columns but not of these ones living, that they were allowed to eat their meal in peace, shows not fine judgment or restraint or generosity, but that the author of the other article was right: there were no survivors. The girls, who'd been eating their rice, hold their chopsticks in mid-air as ash falls upon their heads, and sudden, silence.

The others on the island of Sumbawa ate dry leaves and sold their children for rice (sold to whom? To do what?) They dug up the dead and stole their treasures, which they bartered for food. This would be the long dead, or at least the longer-dead, as the newly dead were left lying on roads and beaches, with nothing in their grasp, except perhaps themselves.

Under the ash now, they've found pottery bits, coins, brassware, and a complete house, with beams and bamboo floors. Inside, iron tools, copper bowls, Chinese porcelain. A skeleton near the hearth. On the porch: a leg, a single vertebra. In another house, a male sits upright, with a ceremonial spear at his side. On his wrist, a bracelet; on his fingers, rings with precious stones; round his neck, a necklace made of brass. And by this other house, a skeleton, lying just outside. His left arm is held up to his head, perhaps against the pumice that is raining down. The deposits of ash are thin, says the writer, so that for the inhabitants the fall of it at first would have been no more than a nuisance, and is probably why they didn't flee. And night falls, and as it does, the families are here, in the dark, the man polishes his spear and someone stands out there, on the porch. Perhaps she looked at the other houses, or brushed a bit of ash from her upturned face. It tasted of nothing, said the sailors on the ship a few days later, as they shovelled out the tonnes that had landed on their deck. They poured it overboard. There was nothing to it but a slightly burnt smell.

*

Aha, but this premise is false, for we assume that because there were no sketches of them then no survivors exist. There's no indication that Europeans, with or without charcoal and specimen bags, lived on the Sangaar peninsula of Sumbawa at all. Forty miles east, there *was* a small British contingent, and from there, in Bima, some four months later, there was famine and deserted villages and the roadside remains of the dead—and the hot August sun, and the Rajah of Sangaar, who'd lost his own daughter to starvation, stood there, holding a gift of rice. Sangaar was twenty-five miles from the summit of Tambora, and the Rajah was in his village that April night. He saw the mountain *like a body of liquid fire extending itself in every direction*. Then the ash, a covering. Then the pumice like walnuts, then the wind, which blew down almost every house and whisked away the roofs, and the roots of the trees, and the trees, and the trees.

And a few days earlier, on Java, where the British Governor Sir Stamford Raffles lived, the sun had been obscured, and there was the sound of thunder, and the native priests there said this dark and that thunder, hear them? Gods, blowing up the mountain. But which mountain? And in the kingdom of Tambora, we don't know what was thought, or if the foreign rule was even felt, or how was it felt, how did it manifest? What did Raffles mean to them? Did the *British contingent* at Bima ever go to Tambora, and if they did, what was said, and what was done, and were they servants to the colonists, were they beaten or did they have to give a portion of their rice and spices to them, and how come here, this writer says that in Tambora and Pekate, just *five or six* survived? And that twenty-six *badly burned people of a party out from Pekate* paddled their canoes away and survived? So: now we picture them, again, these twenty-six, and they aren't frozen under glass or having dinner, but rowing away, their bodies red and blackened and hot, so that if God had stretched out His hand over their canoes, He would have felt the heat rising. The men are rowing, their hands wrapped in cloth, and they're clutching the paddles, or perhaps one man couldn't hold them in his burnt hands and so his paddle slipped away and floated amongst the pumice that grew all around, all around. We know nothing, we never did.

Here is how this writer shows our ignorance: this area was unexplored, by which he means by Europeans, and we know this because Komodo dragons, in the nearby island of Komodo, weren't even discovered until 1911! So reptilian, this accounting. There are twenty-six silhouettes in boats, there are another *five or six* in Tambora and we don't know what happened next, whether they joined the numbers on the roads, whether they died like the Rajah's daughter, what the sky looked like every day after that, whether the road was rimed with dust or ash. That there was wealth is evident by excavations: Chinese porcelain and copper bowls, bracelets and jewelled rings and broken glass. Here he lies, his left hand cradling the back of his head.

And here we too lie, and muddle in half-truths, and put the Rajahs on the wrong roads. We're told that after arrival, Sir

Stamford Raffles prohibited the importation of slaves, but what does this mean in an archipelago of 17,508 islands? The sky is in flames. The bags of rice are not enough.

*

Some months pass. It's December; we're nearing the anniversary of His birth. Did Mary tear? Was she constipated for days? Did she bleed for three weeks after? Did her milk come in on the third day? Was it enough? Did her nipples hurt so much that when He sucked she screamed?

The sheep would have been so warm.

That month, the snow came down in Teramo, Italy. It was red, yellow. Locals took to the streets to placate God.

There is nothing in the record of what they brought on this procession. Perhaps they carried pictures, little icons of Mary and her baby. I picture them holding rosaries and gigantic crosses. I picture them wearing hair-shirts and whipping themselves, red weals hidden by brown shirts. The hairs prickled their skin.

A few weeks later, in Hungary, there was a blizzard that went on for two days and killed 10,000 sheep. The snow was brown, we're told: brown, then *flesh-coloured.* I don't know what that means.

Bread is the body of Christ. His blood is wine. This snow is the colour of someone's skin. This snow is like the largest organ that covers everything else.

Snow covers and is a covering. And there's a hovering in the air. A purple-brown sky, that shade you get at twilight, when the world is in abeyance. It holds its breath: the first flakes fall, gently, inexorably.

Frantic bleats: and the brown snow gathers speed, whirls over heads and into faces, that rapid human blink, those long-lashed, befuddled eyes.

And over here, the steady rhythm of our whips, soft trudge of sheepskin boots, mittened hands and wooden saints.

While over there, floating in a warm sea, a pumice raft three miles across. Like God: large and as indifferent as an iceberg.

She held Him in her arms, and His heart sounded to her ears like a drum, like the far-off breaking of ice. It was hard. So hard, so sad, so beautiful.

The Last Frost Fair

There's ice on the Serpentine, but no trains winding as such around the hills: this is before trains, and what we have now are coaches driven onto paths. Trade gasps.

Men throw snow off their roofs and into the streets—thus are horse and driver vanquished.

This winding lake is thick and solid—how unlike the garters, basking in summer's heat and dust, amongst roses and lettuce beds. Two years from now, after the summer that wasn't, this lake will be chilled, unfrozen, lapping, clear, and Harriet Shelley, pregnant, having penned her note, will walk in.

And this summer we marked the end of the war with France, and the start of the Hanoverian Jubilee. On this Serpentine, the Battle of Trafalgar raged once more, the guns fired on the three-foot-long French ships and we watched them sink as the anthem played. Cakes, acrobats, apple stalls.

Two ladies and a gentleman skate a reel. Between the mountains of snow are ice squares and oblongs, circles. It is winter.

Sometimes there's nothing more beautiful than this. The words "circles of ice," or "snowfall, icicle, mitt," and the way it comes down sideways and straight, how it collects. Here's a fiddle playing against or with this whiteness. Music round and round and oblong, this.

*

They were in the moonlight, reeling to fires, fiddles, tumblers full—and tumblers—ice, and dancing.

I saw these photographic collages, the people of the past building streetcar rails, and they were grey grinning men, their arms bent, hoisting a hammer, or sitting on the edge of the ditch with today's coloured streetcar coming up from behind. Grey flowers grow near green grass and red poppies. A yellow and chrome bicycle by the small-kneed child. There's a woman and her son on one side of the street, and the telephone poles align themselves in grey lines, the woman and her son in grey, array of grey ramshackle houses, and on the other side, a bright tan building, woman with this vibrant hat and shirt.

It makes me sad. We could see it as a continuum, this long long stretch of it, as so our love. But instead the grey ones seem greyer: more dead, more hollow, more frayed, yet also more earnest, sombre, steadfast. More art, and we, with our Condom Shack shops and midriff-baring shirts, seem quicker, slighter: more taglines, less copy. Remember those old posters, the ones advertising flour, where the words go on for the whole page, and the sad housewives are made happy? Though it's an ad, and ridiculous, it still takes time. That mother there is walking slowly, or not walking at all—frozen in a way that the modern ones are not. And so: presumption, of course, but also this ache: I want to connect, to see the ghosts, to walk here and see them at their work. I want the flowers to always have been red, the grass to always have been green, or to be covered in snow and frost, the glitter and shine of ice in the moonlight. What makes me wistful is this invisibility, our unseeingness. How hard it is to see and be seen.

How we long for difference. How this love. How this frozen river brings excitement, reels, wine. How when it was just water, it was a place to get over, and now that it's ice, it's a place to stay. How we celebrate the shake of it. How we laugh. How we look up at this moon. How we sticker the ice with our tents and overpriced gingerbread (the remains of which rest in museums under glass.) How this printing press. How this sheep, cooked on the ice and sold for thrice its value on land. How this gambling tent. How this shack, or that building, the telephone poles, clothing, full-page ads, streetcars. How we conquer with commerce. How I want it to just be us, dancing. How the yard-and-a-half-long icicles, or the shadow of trees upon the snow. How the clouds leave their shadows on the tops of the mountains. How we step on the cloud's shade.

For these ghosts are not somehow better, because they're slower, because they're past.

I'll turn my head for a bit, I'll close my eyes. I'll kiss you here and in every storm—the rain, the snow, the ice: our mouths and the heavens opened. From the ground, the blades of glass, the branches of the very trees are dipped in silver.

*

On the Thames, vast rafts of ice crash against each other, and a few days later, the river's a solid block, with liquor booths, book-stalls, cook-shops, printing presses and a preacher, whose fervour is like to melt the ice. There's so much noise. Theatres are almost forced to close: they are nothing to this. To watch this small sheep burnt over coal fire comes at a cost, not one for the miserly or misanthropic amongst us. Shill out a shilling for a slice of Lapland mutton. Crowds, the pushing worse than a subway's: over here, Bruegel, Bosch, the stumbling drunks. A plumber named Davis walks over the ice with his lead pipes. A Mad Hatter with mercurial eyes is murdering the time.

There are donkeys on the ice, four of them, and, perhaps, an elephant? Gambling tents with women in them, their legs open, the skirts, the roiling sin. Under the ice were waves. Boats dipped in and out again, plunging—the plumber drowns, a child cries, his mouth sticky with gingerbread, all hail and hail and hail the king. Two *genteel-looking* men on an iceberg, too near the edge. Somewhere under all this, their bones are coral-made.

And above, cacophony still? Mouths open and laughing, spittle flying. Gulps of grog. These men were looked for and never found. There will be mourners and horses and handfuls of dirt, twisted ribbons in clenched hands. It will rain. More ice will break, more boats and barges will crash, and there will be pounds and pounds in damages. The crowds will pack the theatres again. Under the ice, it's quiet, and when it's gone, there will be water, more of it. Things will go on. Children will go to bed at the proper time. Lovers will spread their legs in buildings of brick and stone.

*

Transformation: things turn into other things. That water, those animals: thousands of bodies of frozen beasts piled into pyramids at the winter market. Cows, sheep, hogs, fowl, fish: now *stiffened into granite*. All skinned and piled on top, as if, the writer remarks, they were making an effort to climb over the backs of each other, to melt into the woods, to get away: *had an enchanter's wand been instantaneously waved over this sea of animals… they could not have been fixed more decidedly.*

Everything changes. Faithful Johannes speaks and his body turns to stone. He is carried into his master's bedroom; the king looks at him and weeps. "*If only I could have you back to life!*" "*But you can,*" says the statue. "*Just cut off the heads of your two children with your own two hands and rub their blood on me.*"

Workers at the market chop up the animals for buyers, like they are blocks of wood.

And when the thaw comes, you can hear it in the water, loosened, trickling. And when it thaws, do the animals in their pyramids awake and lope off to the woods? Soldiers cut blocks of wine and put them into baskets. As soldiers too transform, men to geometric shapes, lines. This pigeoner bespattered with blood and mud. *His face was smeared with blood from ear to ear; his beard dripped gore; and his clothes were covered with it.*

(Blood from their cut necks are rubbed on stone. *Red lips are not so red / as the stained stones kissed by the English dead*).

It was at the beginning of 1814 that London had its last frost fair. There were women with baskets of hot apples on their heads, and gin tents held up by oars, and people passed out in them.

(There was this: juniper, gin.)

Whereas You J FROST have by Force and Violence taken possession of the RIVER THAMES I hereby give you warning to Quit immediately. A Thaw.

Printed by S Warner
on the ICE

As ginger, flour, egg, transforms. As the stone Johannes comes to life. As the dead children are restored.

Weather, completely.

*

It started with fog that lasted eight days. The lanterns were the briefest of candles. And then two days of snowfall, with a small pause.

Water pipes were frozen and water was needed. They opened the plugs in the streets and a new stream flowed, then froze.

There was ice everywhere and the wind was fierce. In the winter, wrote Hans Christian Andersen, *the windows frost over so strangely, looking like flowers.* And in her kingdom, the snowflakes *came not from the sky but running along the ground, and the closer they came, the bigger they grew. They were the Snow Queen's advance troops. Some looked like big horrid porcupines, others like giant knobs of snakes with writhing heads* [and] *all of them were dazzling white.*

When the first thaw came, it lasted just a day. And with the ebbing of the tide, huge pieces of ice on the Thames rushed downstream. Plates laden with heaped snow formed together and then fought *with great violence,* cracking, crashing, bounding, rising up against each other, *covered with angry foam*. The sound was *equal to the report of artillery.* They were not unlike the pigeons, who *move their unbroken columns like an army of trained soldiers pushing to the front.*

*

I write this during torrents of rain, the wind buffeting the leaves. The wash runs down the hill, over cracked asphalt. *The falls of snow,* I read and thought: how beautiful. The world is green and white and grey, and the snow-in-summer and firewitch dianthus are bright. The cat's ashes are under a small yew, now christened "The Vincent Tree."

I think of words, how we name and the looseness of language. How when my son was a toddler, "bottle" confused him, the shampoo bottle nothing like the bottle he knew. I think of *spirit-dealers* out there on the ice: what deals were they making, with Old Man Frost, with the Snow Queen, and how much were they charging for their services? What currency did the spirits accept? Under the spell of spirits, these people lie in tents, long after the rain started and the warning sounded. It is past midnight and the ice they are on, an *apparently solid mass*, separates. They wake and try to get onto the stationary barges, now crashing beyond their reach.

Imagine waking like this.

Once, I had a deep sadness, and I went outside at three in the morning, and saw this brilliant orange moon, and I took it as a sign. Isn't that how we all see weather, with its sunsets and storms and unsettledness? By February 6th, 1814, the Frost Fair was over. Masses of ice floated down the Thames and thousands of people thronged the banks, disappointed. I think of the swings that were hauled out onto the ice, and the children who must have played on them and maybe adults too, as I did, the other day, alone. I sat on a swing that was roped to a tree, a tree I've come to think of as mine, though the carved initials suggest otherwise. I swung up and every time I swooped back, the leaves on the branch behind me brushed my head. The swing was meant for a smaller child than I, but I wanted to feel I was in this green world, in this air, amongst these leaves. I wanted to be less grounded, so I kept at it.

*

What I'm getting at also is the coincidence of time. Let us refresh. 1810: Alexander Wilson is in Kentucky, inside a cabin. He hears *a loud rushing roar, succeeded by instant darkness.* A tornado, he thinks, but his hosts correct. He describes the birds thus: *glittery undulations.* 1813, John James Audubon writes about Ohio, Kentucky, the pigeons whose numbers won't decrease, the storm of them, the *hard gale.* 1814: London's last Frost Fair, with Old Man Frost and Frostiana, and the lives lost, and the Serpentine, covered in ice. 1815: Tambora, another transformation. 1816: the year there was no summer.

Birds are storms; storms are gods, spirits, drunken and invoked.

Here's both meaning and coincidence. It's true that the Frozen Thames is connected to the summer of two years later, that this Little Ice Age didn't help matters, that the ash that travelled up merged with ash, that this part of Earth was already pretty damn cold.

This world feels elegiac. The last Frost Fair, and a year later a volcano, and no one skating knew that, no one knew it would be the last one, that bridges and embankments would make the Thames flow faster: it would never freeze like this again.

No one knew (though some did warn) that the birds eclipsing the sun wouldn't exist a hundred years later, and that those men in such formations wouldn't be the same again. No one knew that this piece of gingerbread would end up in a museum and that descendants would stare at it through glass.

If there is wisdom, it's nothing I know. It's all just birds and storms and hauntings. We look behind and scoff, as if those ahead aren't doing the same.

The Garden of the Fugitives

Ashes were already falling... I looked round: a dense black cloud was coming up behind us, spreading over the earth like a flood. "Let us leave the road while we can still see," I said, "Or we shall be knocked down and trampled underfoot in the dark by the crowd behind."

I saw a bowl of olives and another of figs, set out to be eaten and now carbonized, preserved, for oh, these thousands of years.

I saw the cast of a dog, chained to its post: the twisted body, front paws in the air, head curled by hind legs. I saw men with pickaxes, the cast of the man with his head upon his knees, the cast of the woman, far from her child who lay on his belly, his arms above his head. I wanted to go to him, I wanted her just a few feet closer, her arm there, there.

We had scarcely sat down to rest when darkness fell, not the dark of a moonless or cloudy night, but as if the lamp had been put out in a closed room.

The mother's head is up, she's looking at her son, but he doesn't see, he has his head down, he's already dead. Maybe she is too. It's so hard to know. It's so hard to watch this, to stand and stare at them through glass.

You could hear the shrieks of women, the wailing of infants, and the shouting of men; some were calling their parents, others their children or their wives, trying to recognize them by their voices.

And yet I'm not sure it is a mother, or even a woman, and whether they were precisely that distance and direction apart. I don't know if she was the one who put out that bowl of figs and maybe he grabbed a couple and she yelled at him, before the lamp was put out in this closed room.

Many besought the aid of the gods, but still more imagined there were no gods left, and that the universe was plunged into eternal darkness…

All I know is they are not touching. This hurts me more than anything.

Ragnarok

. . . Wolf-time, wind-time, axe-time,
sword-time, shields-high time, as the world
shatters and no one is spared by anyone.

The Great Winter came, presaging Ragnarok. Three years of snow and storm, the snow constantly falling from every point of the compass.

Birds dead in the field, *a river full of knives*. Shorn sheep shiver in the summer snow, then stop. Creatures in the dark. Hearts in the wet.

The wolves catch up to the sun and moon—this bright sun blackened, this *bright sun extinguish'd*. The sun *black as sack-cloth of hair*. Moonless, perpetual night: lanterns swinging blindly over the blind.

I hid my eyes and wept, as *every mountain and island were moved out of their places*. I shrieked, and the dragon Nidhogg looked up before going back to sucking on the dead.

The Earth sinks into sea—and behold, this ship, with Loki at the helm. A ship full of fever—manned by ghosts and built from the nails of the dead.

I will kill her children with death.

When the survivors got off, they wavered like flames and then fell onto the streets, their eye pits filling with snow.

*

Seasonless, herbless, treeless, manless, lifeless: this is the Earth, great swathes of it.

In China, they sell their children, suck on white clay, convert their dead fields into poppy dens.

In Ireland, lice destroy thousands. In India, the monsoon comes late, the trade winds fade, thousands of painted ships lie upon the painted sea. When the rains come, they flood into choleric mouths.

Eat cats, eat moss, eat dirt:

That God has expressed His displeasure towards the inhabitants of the earth by withholding the ordinary rains and sunshine cannot be reasonably doubted.

Hailstorms, a fever epidemic across the frontier, a great comet, an eclipse, vast flocks of pigeons in the skies. Autumn: swarms of squirrels like a mass of roaches head south and drown in the Ohio River. Winter: Earthquake—the land swallowed, the land dark, the trees. Entire islands vanish, the Mississippi flows backwards, church bells ringing in every lightning storm—

but the demons are not afraid.

*

The Frost Giants come closer. They pelt the earth with snow and ice. The men fight for food. The hall of Hel is filled with all those who starved, hollow-eyed, their hands out. When alive, they were mistaken for an army on the march.

*

But in those days too, men sought death and could not find it. They desired to die, yet death fled from them.

For who would want to live, with all whom you love dead?

*

Great is the distress, Oh Lord, have pity.

Wherever one god falls, up springs another, amidst the wind, the hail, the snow, the ice. This man here, these cults there, families moving to the Holy Land to escape the coming plague, an injunction not to bury their dead. *Let the dead bury their dead: but go thou and preach the kingdom of God.*

Flies and buzzards surround.

I screamed, and all about me were men and women keening, kneeling on the dark earth, their hands raised to the sky.

The great day of his wrath is come; and who shall be able to stand?

Not even these. The sea rages. The stars rip from the sky.
Gods fighting gods, gods living, gods dead.

Blood on every ear of corn.

Here's the weather, there's the war, here's the weathered dead.

Daedalus

What I didn't say was how the mud wasn't really mud. It was wet earth made of dirt and empty shells, iron scraps and organic waste. What I didn't say was that the arms and legs of men stuck out from it: *everywhere I look, bodies emerge, shapelessly, from their shroud of mud*. What I didn't say was that it was the worst rainfall in thirty years. What I didn't say was the smell. What I didn't say was that water was poisoned by the bodies and excrement that had sunk into the mud. It travelled in black waters across the plain. I didn't talk about that journey. I didn't talk about the clouds of smoke that wavered over the mud. I didn't mention how for years after, nothing but weeds could grow.

I didn't talk about how the Tank Corps learned from locals that the rain and bombardment would wreck the watercourse system. The battlefield would become a swamp: *no more unsuitable spot could have been discovered.* They got no reply. I didn't say that they forwarded charts, showed where the larger pools of water would collect. The note back: *Send us no more of these ridiculous maps.*

I didn't talk about how the tanks bogged down in the mud. I didn't say how seventeen were abandoned in what became *Tank Graveyard* and how when men tried to get them out, it became theirs too. I didn't talk about how the tanks crawled one mile an hour.

I hate the idea of thrusting an Army into such a daedalus of mud and water. And Haig responded: if they advanced just *a bit, it was all to the good.*

I should talk about that journey, about the water flowing over the plain. There are so many words for water. Brook, stream, rill, beek, eddy, pool, riffle, tributary, torrent, streamlet, reach, outfall. Outfall. The hard-edged ones like pumice stone.

A *daedalus* being named after that which Daedalus created. Not the later invention.

They did advance a bit. I mentioned that. In two months, the largest advance from the Salient was three miles. After four months, in just a few days, the Germans won it all back.

And what was at the heart of this labyrinth? Just more slime, and perhaps Haig, listening to his God, and commenting *all to the good, all to the good.*

While men and this waste seeped into beek, eddy, rill. Another and another: bile, splutter. Sixty pounds on each man's back and mud makes more. Hands reach up. Open mouth. Open mouth. An O – this blown-out candle and broken thread.

*

Half man, half animal. We were always thus.

It was dark, sky lit with flares. The road had an inch of slime and then you had to turn and there were just duckboards, and then the duckboards were gone, and you had to go off into the mud to get to the next one, and men, blind, stepped on drowning men's fingers and kept trudging, and mules loaded down with shells were stuck and struck: the eyes, the eyes.

Men swallowed by this dark and clutching earth. Though "swallow" is not right: too complete an action, no evidence, and yes, let's remember the evidence, the man's hands upon your boot and lace. A tambour could not record it thus, these halfway disappeared and gleaming.

And is this earth a mouth to swallow or a hand to clutch? Perhaps it's right that in writing about this morass, there should come a metaphor so mixed. And that regardless, it is as bodies, maybe giant ones, maybe just one, as Gaia in that book I had as a child, her shoulders and breasts green hills and the sky Uranus looking down at her, his eyes twinkling stars. So this plain, and broken trees. And Cronus, but his children are forever lost, swallowed, then thrown-up by plough and axe, remaining and unremained. *Remains resting at the Bell Funeral Home* I read once, in the paper, in the death announcements for my great-grandmother, and I was puzzled at first about how she could still be resting, but all that was meant was that who she'd become was at the funeral home. *She* was the remains. Of herself, I suppose.

These men remain resting. The boot comes down on the white fingers. On the black.

The dark labyrinth, and the minotaur. *There came, with time, a sort of deadening.*

A swallow, above, wings flapping.

*

Procne: woman turned swallow. Her son, Itys, flung his hands around her neck and gave a kiss: *"Oh darling mother, I love you so much!"* She pried his hands away and dragged him into another room: he stretched his arms to hug and: *"mother!"* She stabbed him in the chest and cut his throat, and then she and her sister (oh, nightingale!) tore off his limbs and threw him in the cooking pot. Her husband ate him and called for more.

Her arms became wings and the feathers on her breast were red as blood. And rapist Tereus turned into a bird *armed for battle*. The hoopoe is a thief, they say, and in Scandinavia, he's the harbinger of war. In Estonia, his cry means death.

And these are they which ye shall have in abomination among the fowls; they shall not be eaten, they are an abomination: the eagle and the ossifrage, and the ospray, And the vulture, and the kite after his kind; Every raven after his kind; And the owl, and the night hawk, and the cuckow, and the hawk after his kind, And the little owl, and the cormorant, and the great owl, And the swan, and the pelican, and the gier eagle, And the stork, the heron after her kind, and the hoopoe, and the bat.

"Oop-oop-oop," he calls, and his battle cry is this: a croaking like frog, a hissing like snake.

And what of the swallow? Once they were thought to replace war pigeons, but the messages got lost. They perch out of reach, they alight on roofs or dead trees and look down at this field of smoke and mud. They foretell land; they're luck to sailors. A group of them is called a sweep.

But Itys, his hands around his mother's neck, is cut into a pot: an *armless, boneless, chickenless egg.*

*

Now Perdix, Daedalus' nephew, walked along the sands and came across a fish's skeleton. He felt its spine, the bones rasped across his hand, and thus he invented the handsaw.

Perdix, who then joined iron with iron to make the perfect compass circle, as bombs did later on these fields. Or in this caldera left by the gods, a circle without beginning or end, a wedding ring.

Perdix, thrown off the cliff by his jealous uncle, and transformed in mid-air to a partridge, flying low and nesting close to the earth.

What is there to say about the earth? It fills mouths and nostrils, and sticks to eyes like sleep.

I want to hold onto that moment when he walked along the sands and saw the fish. It was a clear day, this soft breeze, like the summer that came before the war. Everyone talks about the weather that June, how warm, how plentiful the picnics, how white the wicker basket under trees.

How the fish was picked clean, how bones, how fresh the waves upon the strand. The gulls were just that, swooping, and this spine could be a comb against your hair: you draw it through the windblown curls. Death comes to the old, to the swooping gull and the flapping fish, the sleeping man, newspaper folded on his chest.

There's tea with lemon and lumps of sugar, and a young boy, this open space: a beach, a marsh, an impossibly huge cerulean sky—he wandered barefoot, his trousers rolled, and there in the distance: a glint of bone. As if the world was ever this innocent.

And yet, it was, it was. No trenches, a labyrinth not yet built.

*

My son drew a picture and described it. "There is a cave with some sparkly rocks and blowed trumpets that sounds like a flute to make music from both sides." It was autumn. The leaves that surrounded the house were yellow and crimson and orange, and when we walked down the path he bent down and picked only the red ones. He was going to start a club, he said, and it would be called "The Glories of Nature" and he would be the boss, he would bring in the glories of nature, they would be his.

Look, here's another one! Look at *this* Glory of Nature! We raced our glories, "look look, mine's winning!" while the other got trapped in the stream with sticks and leaves.

There were blowed trumpets and in Flanders, they thought the world was ending. It's the year of no summer and the trumpets blowed for evening, but when the trumpets blowed the people ran out and fell to their knees. Where is your God now? And Jacob Flanders, on the beach, years before the war, *sobbing, but absent-mindedly,* runs with a whole skull, *perhaps a cow's skull, a skull, perhaps, with the teeth in it.*

There's this way and that. *The war tainted the past… revealing and making explicit a violence… latent in the preceding peace.*

Eighty years on, this sense of crouched and gathering violence has been all but... filtered out.

So for us, the tea, the fish, the old man, parasols. For Woolf, the young choirboys of '06, whose *great boots march under gowns...* and when Jacob drinks with the rustics, there's *Old Jevons with one eye gone, and his clothes the colour of mud, his bag over his back, and his brains laid feet down in earth among the violet roots.*

It's peacetime and sunny, and Sassoon hunts foxes.

Now the sky. But the music came from both sides. But the music, but there was music, but, surely, music. Flutes not trumpets. Sparkly rocks on the tops of dark caves.

The Great War was *the turning-point in the history of the earth,* says Wyndham Lewis, and I think of the farmers turning the earth, what they found there. The iron harvest. The sky *sullen grey and the trees of black iron.* Tambora too turned the earth; there are always points. But not like this.

Perdix, walking, picks up fish spines, and Jacob running, cradles a skull in his arms. And my son picks leaves. My son has captured the Glories of Nature.

*

Above the *dead sea of mud,* where *skulls appeared like mushrooms,* there are larks at stand-to in the morning and nightingales bombarding night.

We heard the guns. There was continuous fire. It was a background to the singing of the nightingales. It has often seemed to me that gunfire makes birds sing.

Sweet Philomel sings after slaughtering her nephew. There are dead nephews, dead sons. He *slew his son, / And half the seed of Europe, one by one.*

Daedalus was expelled from Athens and ended up in Crete, where he bred a monster and made the labyrinth, wherein the young men wander.

The trenches are a labyrinth, I have already lost myself repeatedly… you can't get out of them and walk about the country or see anything at all but two muddy walls on each side of you…

There are the skulls cropping up like mushrooms, sticks and crimson leaves, and a child rolling down the hill, laughing. The yellow jacket, the yellowjackets, the larks. *I used to hear the larks singing soon after we stood-to about dawn… they made me more sad than anything else.*

He crouched down in the maze of his invention.

*

Later, when Passchendaele was over, Haig's Chief of Staff was driven to the front. The story goes that as he viewed the mud, he burst into tears: "*Good God, did we really send men to fight in that?*"

"*It's worse*," the man beside him said, "*further on up*."

There's a way out of here, Daedalus told his son, and so they made wings.

There's only one way out, and it means not being human. Go now, up beyond the rain clouds. Go quick then, my darling boy, my love.

Medusa

I

Things turn into other things.

In fairy tales, metamorphosis is transitory. Faithful Johannes comes back from stone, the swans become men, everyone in the castle awakes. Aslan breathes on a lion and the gold spreads as a *flame licks all over a bit of paper.* All is turned back, from *deadly white* to a glorious *blaze of colour*.

In nature too, transformation never holds. The leaves fall, the snow descends. We battle the marsh, the flame, the wind, the ice. We hold back the sea with a faltering finger. The trees are a column of soldiers against the wind: the wind holds fierce, the wood breaks.

In Greek myths, the transformation stays. Daphne, a laurel tree, forever green. Perdix, a partridge in perpetuity. Birds, spiders, Actaeon a mauled deer, dead and still dead: transmogrifying back in the underworld, perhaps, or among the stars, but not here. Arachnids on the branches or by the woodpile, where we have blocked out the wilderness. Yet behold this corner, here.

I want to believe in true transformation, but I know this sandbag is shot, this seawall temporary. These grasses clutching the dunes are not enough. I know there are hearts and thoughts that break and mend and break again, that as Faulkner said, the past is not past, that what you put your trust into might be rocks, but that rocks were and will be sand.

What I believe in, more than Daphne and Perdix, or even the fairy tales where the good are fully restored, are werewolves. A people who are mostly good, except when, terrifyingly, they're not. I have it in me to believe in a goodness that happens most days of the month, though history proves otherwise. I choose not to see this. I minimize the evil or make it predictable with clear warnings, a beacon of soft light.

Think of Medusa, not the woman made forever monster, with snakes for curls, and a look that turns men to stones no god will ever breathe life upon. Nor even the jellyfish with its sting, which strikes the living dead. Think of the shipwreck, with its raft immortalized by Géricault. And what I am struck with here is how soon we turn, how little time it takes.

*

My son is a werewolf, he says, but one that comes out every Friday night at midnight, not only during the full moon. He isn't a bad werewolf, but a good one, getting rid of evil monsters, but sometimes he's bad too and if he says the word, he'll turn into one right away. So he's running through the house, and in the backyard, where the garden is full of wildflowers, forget-me-nots and lilies of the valley, and he's whispering about being a W.W. because it's Monday, he's not ready to turn yet.

Werewolf in Europe. Weretiger in India. Werehyena in parts of Africa. Whoever attacks, runs back to the hills with your child in his mouth.

Of all the teeming perils of the night and the forest, ghosts, hobgoblins, ogres that grill babies upon gridirons, witches that fatten their captives in cages for cannibal tables, the wolf is the worst for he cannot listen to reason.

*

We have not blocked the wilderness. Here, here, in his yellow eyes and the howling that rattles the panes. In April, spots are seen on the surface of the sun, like *a spider, having parts extending from the main body.* Soon the sun will be obscured and the whole world will be in darkness.

The sun is ill and the moon is dying and life on earth will end July 18th. The prophet is thrown in jail, but some still hear and crowd the churches or kill themselves. A cook in London with a rope around her neck. Her feet dangling, hailstones smashing the glass, and rattling, so, in her throat.

And now from the woods the wolves come, lean and grey, their eyes gleaming from the light of the fires we lit this summer to stay warm. That August, in rainless Carolina, the meadows are white with frost, the dew congealed to ice. But the frost *killed nothing, as all was dead before.*

Farther north, the men have guns. The men are not wolves. There is a way of seeing a werewolf, even before he turns. "*Never stray from the path,*" intones the grandmother. "*Never eat a windfall apple, never trust a man whose eyebrows meet in the middle."* Low-set ears, curved fingernails, a way of walking, a swing and a lope.

You try and keep this fear close in your heart, the sealed lips, but out it seeps, like smoke, like the forest fires that summer in New England. The smoke so thick that *though there are no clouds, the sun is not to be seen.* The sky is black, fences turned to charcoal. Outside Boston, the wind blows cinders onto vessels. Ships wreck, and rack, mouth open in pain:

We stand aside, separate, as though we're not this. Wolfish, a wilderness.

*

The living snatched souvenirs from the dead. It was still dark on Waterloo when the bodies were stripped, a field of helmets. 100,000 horses skinned and unshod within the first twenty-four hours. In April of 1816, there was still looting, people digging up corpses, taking anything they found in pockets. Books, bullets, helmets, hats: there were buyers clamouring, Lord Byron among them.

Teeth were wrenched from the dead to make dentures for the rich. The amount from Waterloo was enough to ship supplies all around Europe and even across the Atlantic. Remains not resting. Mouth bloody and broke, eyes blank. Smile! What a picture. Daguerreotype *de la guerre*.

*

If you are bitten. If you rub your body with a magic salve. If you wear a belt made of wolf skin. If you sleep outside on a summer's night, with the full moon shining on your face. If you drink water from the footprint of a wolf.

And, once transformed, what to do but to kill the sheep in the fold, the children in the bed? To rut amongst the bedclothes with the women, pawing, licking; to fuck wolves from behind.

What to do then but summon, as Moeris did, the dead from the depths of the grave, *and move the standing corn into other men's fields?* To cause an eclipse or a hailstorm, which in half an hour *had very strangely ruined all the fruit of that country.*

In some countries, you must, after killing a werewolf, cut off its head with a spade, and get it exorcised by the parish priest. Otherwise the corpse will transform into a wolf, prowling the battlefields and drinking the blood of those dying men.

Picture him here, a grey wolf slinking along in the shadows while all about him are men cutting off the buttons and boots of the dead.

*

And was it a werewolf then who caused the waters to flood, the thunder to rage, the hailstones to pelt, the land to slide, the ice to form, the trees to fall, the houses to slip, the cattle to drown, the people to starve?

A dark spot on the sun.

The Devil, says *Malleus Maleficarum*, cannot truly change men into beasts, for only God can create and alter matter. But Satan can, *by moving the inner perceptions and humours, effect changes in the actions and faculties... Wherefore S. Augustine, speaking of witches, says: "These are they who, with the permission of God, stir up the elements, and confuse the minds of those who do not trust in God."*

*

"Panic": the state caused by hearing goat-footed Pan rustle the bushes as you walk these dark woods. Mud, volcano, wolves digging out corpses, the panic of war. We're hearing Pan howl his despair. We're running through brush, twigs snapping, the thornbush catching our clothes.

"Who goes there?" There is a photograph of a tree growing through a piano. A Ferris wheel covered in leaves. An abandoned mining town with hillocks of sand in the house. This lovely loneliness of the vanquished, the near-forgotten. You can see this with the battlefields of World War I. Where the trenches were, a slight dip in earth. It starts now. Heinrich Böll, on the bombardment of Cologne in World War II, writes:

...It was a question of botany. This heap of rubble was bare, naked, all rough stones and recently shattered masonry... with not a blade of grass in sight, whereas elsewhere trees were already growing, pretty little trees springing up in bedrooms and kitchens.

All these halflings, piano trees, a town of Chernobyl, part-city, part-forest. The monsters, too: Pan, Minotaur, Medusa, Lycanthrope, the biggest danger someone who is part us.

(A reef, a raft, a white tongue. Ship full of weapons and wine).

We are our own gods.

II

As soon as it was clear the ship had foundered, soldiers and sailors got drunk, ransacked the trunks and strongboxes. They searched for fine dresses, precious objects. Snatched plates, candlesticks. Broke into the captain's cabin, guzzled wine and booze. The men wore jewellery, mink stoles, ladies' dresses: layers upon layers, as much as their bodies could hold.

The world pitches, the floor we stand on starts sinking. And when the floor starts sinking, we do the instinctual. Mean, grasping, brimming with hope.

*

Howl, howl, howl, howl: in the middle ages, peasants called famine "the wolf."

"Scritch, scratch" at the door with his sharp nails, and then "ow-ow-awroo!" The thump as the last junk of wood falls from the fire. A scant pinch of flour in the bin.

A German rhyme reads: *at eleven come the wolves, at twelve the tombs of the dead open.*

1816: The population of France swells with demobbed soldiers. French factories that had flourished with the wartime blockade on British goods collapse beneath the avalanche of cheap imports. The French linen industry is in ruins. In Belgium, destitute workers burn mountains of English textiles. The price of bread rises, and there's rioting in the south of France. On the west coast, peasants armed with pitchforks and sticks stop a shipment of wheat, steal as much as they can.

And it's not much better in England, where employment on the London docks goes from 1,500 to 500. In West Dorset, the price of wheat rises fifty percent in just a few months. People smash shop windows and break into mills, carrying off sacks of flour. Others wave flags: *Bread or Blood*!

It rains. And rains. And rains.

Residents of Grenoble, France are trapped between two flooding rivers. The Seine continues to rise. That summer, the church orders nine days of prayer for better weather.

There's famine, the food is disappearing, it's gone, "ow-ow-aw-roo!" "Husha, husha!" whispers the Madonna in the rocking chair, but is there a chair left in this story? Perhaps it's burning now and she's on the floor, getting some heat in this summer that is no summer, and the baby keeps sucking, but it's getting harder, the milk is drying up. Outside the rain falls and keeps falling,

on the Western Front, the rain falls, keeps falling,

but right now, we're en route to Senegal, we're in the tropics. We're on this frigate, part of a convoy that will take us to Saint-Louis to re-establish the French colony that was Britain's during the war. The ship hits a reef, there is screaming and, almost immediately, looting, the soldiers are getting drunk—

Let's take a break, look up facts about Saint-Louis, a major port for gold, leather, gum arabic, ambergris, and slaves.

Killing whales and seizing men—here are the ruins of a slave holding cell. Easy enough to say they stopped being human, got uncivilized when they broke into the captain's cabin and drank all that rum,

but we were wearing this perfume and opening the tombs of the dead long before.

*

Still, the turn is distinct, the moon has risen, the light is on your face. You've drunk the wine, the water from the wolf's paw.

It was a calm day and the frigate was sailing too close to shore, in an area riddled with rocks. The water had changed from deep blue to soft green. Sand was swirling in the waves; kelp floated on the surface.

Both the ensign of the watch and a colonist who'd travelled this route two times before told the captain that they were sailing straight for the reef. They weren't believed. Soundings were taken every half hour: Eighteen fathoms; ten; six. 16 feet, 23 inches—July 2, 1816: a calm, clear day.

And then it was night, and there wasn't enough room in the lifeboats and so they built the raft. On July 4th, waves struck the hull of the frigate, and the sound was like thousands of pigeons' wings. The keel split in two. Soldiers, who were coming to repossess a land that should never have been theirs, swayed on a broken floor in the middle of the ocean and grabbed plates and candlesticks while others stood in strict formation with guns, threatened to shoot anyone who tried to leave.

And there were others in the darkness, in this storm, with the lightning and the waves crashing, children and mothers and fathers, leaving France and famine behind—

while elsewhere on the ocean bobbed the ships. *Emilia, Voadora, Eleonora, Rosa, Aurora, Amistad, Empresa, Luxuria*— their destinations Amazonia, Martinique, Bahamas, Bahia, Cuba, Brazil. From France, in 1815, 1816, we have *Flore, Africain, Actif, Hermone, Petite Louise*, *Jenny, Belle, Bonne Mère.* Bonne, Bonne Mère.

The captain of the *Petite Louise* was Lancelot.

The *Africain* left Saint Malo, France on August 27th, 1815, and began trading in Gabon exactly two months later. In the New Year, she left Africa, and on March 7th, 1816, arrived in Martinique with 263 slaves, having lost fifty-six en route. *Voyage length, home port to slaves landing (days): 193. Middle passage (days): 49.*

On July 23rd, 1816, six days after the survivors on the raft are rescued, the *Africain* comes home with goods in its hold.

*

For it is not enough for me to just tell one story, as if there is this one thing, this one wreck on the stormy sea. To them, on those boards, it must have seemed so, the raft must have been their only world, and yet I? I have the eyes of God.

I picked up my son from daycare and he said to me, "Mommy?" And I said, "yes?" And he said, "Today I had a hard time staying human." When we got home, he stayed outside, crawling on two deck chairs and howling like a wolf.

Claude Prieur, *Dialogue de la lycanthropy*: *There is no savage beast wilder than man if he is left to himself.* Or, proverbially, this: *Man is a wolf to man.*

Yet I think the wolf frightens primarily *because* he is not us. I think also that we're scared of us, our own wildness. It is a math problem with two different answers. It is, it is not.

I look out and see the ships and the pigeons, men waving pitchforks, transforming into wolves. Mount Tambora erupts—and if I have the eyes of God, it is of an uncomprehending deity's, who stares and stares and cannot see.

*

By July 5th, 1816, the frigate had taken in too much water, and so at 5 a.m. there began the evacuation. After forty men clambered aboard the raft, it sank a few feet in some places, so provisions were rolled off into the sea, leaving just six tubs of wine and two containers of water. Someone from the frigate tossed them a twenty-five-pound sack of biscuits and missed. It landed in the sea, where it was retrieved, a soggy, salty mess.

Sailors swam to a barge that could have taken fifteen more people. They pleaded to be let on, but were met with a sabre, and frantically swam back to the broken ship.

At 11 that morning, under the orders of Governor Schmaltz, a man in the barge grabbed his hatchet and cut the tow ropes attaching the raft to the barge. They were on their own. There were no sails, no oars, no ropes, no charts, no compass, no anchor, no rudder. There were wine and weapons, some sodden biscuits, 146 men and one woman. A butcher, a baker, a master cannoneer, a barrel maker, servants of staff officers from the *Medusa*, an armorer, members from the Philanthropic Society, soldiers from the Africa Battalion and just twenty sailors. The leaders huddled in the most stable part of the raft, the deck in the centre.

A storm that night: clouds, darkness, waves smashing against the raft. Hold fast to the ropes. Cries, shrieks, bodies overboard and back, over and back, bodies smashing into the raft, bodies trapped underneath the masts and spars, salt spray in their wounds. *It was impossible,* writes Julian Barnes, *to form an idea of that first night which was not below the truth.*

*

Morning, July 6th. Pre-ration roll call: about a dozen dead from wounds or from the sea. A baker and his two apprentices cannot stand the heat and thirst and throw themselves in. Thirteen, fourteen, fifteen.

Night. Hallucinations. Storm worse than the night before. When the sea calms, a group of soldiers break a hole into a wine cask, drink the sea water that rushes in, and the wine. Crazed with drink, one man raises his hatchet and begins to cut the cords of the raft. He is stopped by an officer, run through with a sabre, and thrown overboard. At this, the drunken soldiers, clutching swords and bayonets, rush to the centre where the officers are. Those without weapons bite their victims. One soldier chomps at another man's Achilles' tendon while three collaborators attack him with knives and the butt of a rifle.

Is the moon full? Can you glimpse it behind the clouds? It is the second night.

Achilles was almost immortal. You couldn't call him a halfling, exactly, as he was more than half a god, sulking in his tent, going to war, killing so many that the river was choked with bodies. Treat my corpse with respect, pleads Hector, and Achilles responds: I will "*hack your flesh away and eat you raw.*"

*

The rage, the rain that's falling, a dead man in the dust. I raise my head. *The wolf is the worst for it cannot listen to reason.* The men are biting each other. It is the second night.

There are cries for bread, for a chicken leg, for boats to save them. Some believe they're still on this ship, named for a monster. I don't know how to write this. The sun rises on the third day. July 7th is my son's birthday. He was born in the wee hours of the morning, when the darkness thinned and the living found sixty dead. Last night, in the fighting, two barrels of water and two barrels of wine were slung into the sea. It was the third day and there was just one cask of wine left. I could be a God, rising up in anger at Haig, or at Achilles, who blocks the waterways with his dead. His name can be interpreted to mean Grief. Grief of the nation, Grief of the people.

People begin to eat their hats or the scabbards of their swords. And it is on this third day that people start on the dead that surround them. Stomach, shoulder, thigh.

The weather was calmer that night. Men dozed standing up, water up to their knees.

*

Day four, July 8th. On this day, there is frost on crops all the way from Maine to Virginia. Looking back, some Americans will call this summer "The Mackerel Year," after all the mackerel New England farmers had to feed to their pigs.

Also on this day, in the heat, *ten or twelve* more are found dead on the raft. All but one, saved for food, are consigned to the deep.

Ten or twelve. It is the vagueness of the number that gets me, just as *five or six* did in Tambora. Which is it, and who are they? What if it were eleven?

*

Day four, July 8th, another riot breaks out. Men scale the mast to get money that's hung up there for safekeeping. More blood, stabbings. Dying men carrying candlesticks, holding a plate.

One, two. Why, then, 'tis time to do 't.

Tally thus far: 117.

*

Day five, July 9th. There is enough wine for everyone for four days.

Day seven, July 11th. At 9 p.m., trumpets sound during a thunderstorm in Ghent and *suddenly cries, groans, tears, lamentations, were heard on every side… It was not without infinite trouble that the cause of this extraordinary terror was discovered.*

That day, two soldiers sneak behind the remaining wine cask, drill a hole, and start drinking. For this crime, the men are executed at once. A twelve-year-old cabin boy dies in the midshipman's arms.

Easier now just to count the living: Thirty, twenty-nine, twenty-eight, twenty-seven.

They kill the weakest, so as to preserve enough wine for the rest. Among them, the one woman, who was thrown overboard during the first riot, and rescued, then thrown overboard in the second riot and rescued. She is killed alongside her husband, who was badly wounded in the head. She was a sutler, the record shows, selling supplies to the military.

Twenty-six, twenty-five, twenty-four, twenty-three, twenty-two, twenty-one, twenty, nineteen, eighteen, seventeen, sixteen.

Fifteen.

Fifteen left. A white butterfly lands, a sign that land is near. Another, and another. Then a gull, another, another. The men pray for a storm to send them close to land.

Day eleven, July 15th. Sharks surround the raft, but do not attack. Instead, the men are stung by medusae, the Portuguese man-of-war, whose filaments are like a Gorgon's hair.

Day thirteen, July 17th: a sail appears, then disappears. A few hours later, the *Argus* sails towards them, and her men help the survivors aboard.

July 18th: the day the Italian prophet forecasted would mark the end of the world.

July 19th: the *Argus* (a giant with a hundred eyes? A son of Zeus, a builder of the ship *Argo*?) sails into Saint-Louis, drops anchor at 3 p.m.

*

In 1815, just a year earlier, the French slaver, the *Petit Schooner de Nantes*, left Saint-Louis with 179 slaves. It was captured shortly thereafter by the British. The record reads: *Outcome of goal for owner: original goal thwarted (human agency).* For other ships, there is *Voyage completed as intended*, or *Outcome of goal for owner: thwarted (natural hazard).* Between embarking in Saint-Louis and landing in Sierra Leone, *Petit Schooner* loses twenty slaves. It is difficult to find out what happened to the 159 who survived.

I'd like to think they made it back home. I'd like to think there's no danger when the moon wanes. I'd like to think that it takes thirty days, not two, for us to bite.

In Brittany, it is thought that the souls of the dead are carried to islands in boats. On a calm day, you can hear the sound of the oars as they cut through the waters.

There is a white haze that summer, and over there, a sooty darkness. Either way, you cannot see. All you can do is hear the oars dip in, the brief silence as they slip out. If you squint hard, maybe you'll spy white shadows hovering, shimmering over the black boats.

The Children of Famine

Once upon a time, there was a woman who had two daughters, as beautiful as sunshine on snowfall and as lovely as unplucked plums.

Then came the rains. For weeks that summer, it rained. The Saale River flooded. You could see only the top of the bridge: below, cows, legs splayed open, mouths filled with mud.

"Look, *Mutter*!" said the youngest: *Look at that man on that hill, or sitting alone in his fishing-boat, rowing across the fields where he recently guided his ploughshare.*

And her mother replied: Yes! And see, see that one, *...sailing above his cornfields or over the roof of his vanished farmhouse, or casting his line in the top of an elm tree.*

It was as good as a play.

*

They ate rotting grain, rotting potatoes, rotting rats. They boiled weeds called *pigs' ears*, chewed bread made of sawdust and straw, gnawed on their pet dog. Rain dripped down the eldest daughter's face when she went to sharpen the knife.

In Wüerttemberg, cadavers trudged through the streets and flocks of children cried out for bread. There were 3,000 more deaths than births. The children perched above your head, branches black and dead.

1817 was The Year of the Beggar. *Das Hungerjahr.*

*

Once upon a time, the two daughters had in their four white hands a short length of string. They were trying to play cats' cradle with it, but their fingers were numb with cold. And then one of them, the youngest, grabbed the string from around her sister's pinkie, and started to chew, just to have something in her mouth.

Their mother stirred herself from beside the dead fire. "*I've got to kill you,*" she whispered to her girls, "*so I can have something to eat.*"

The eldest daughter looked up from her sister, chewing on string. "Oh no, dear mother. Spare me, and I'll go out and find us something."

So she opened the window and swam out to see the world. She swam and swam, around farmhouses and church spires, but nowhere could she find food, for all the grain and animals had been drowned long before. Her thin arms were tired now, and she badly needed to rest. Finally, in Laichingen, she found a hill that was soaked, but still above the waters, and there, the rich were standing. She clambered up to the top, dripping and covered in topsoil, the farmlands of her friends. "Please!" she cried, but the rich could not hear her. They were jostling for property, they were buying up the homes of the poor, and their noise on the hill was like a million million hailstones falling onto cobblestone streets.

The eldest daughter cried out again, and finally one man heard her, and gave her a little piece of bread. Or not gave: paid her for services. She dove back down with the bread in her bruised mouth.

She swam for days and when she came home she cried out, "Here, here you go, Mother, youngest daughter" and as she did so, the bread fell from her mouth. Her mother and sister scrambled to eat it from the floor. A small crumb was left for the eldest daughter.

A day passed, but the bread was not enough to still their hunger. So the mother said to the youngest daughter, "*Now it's your turn to die.*"

The youngest daughter looked up from the floor. "Oh no, dear mother. Spare me, and I'll go out and find us something."

So she crawled up the chimney and jumped from rooftop to rooftop, her small white legs making large leaps. She leapt and leapt for hours, hours, days, until she could see land again, houses and churches and cobblestone streets. Then she slid down the roof and landed amongst thousands of men and women, ripping chunks out of a living chestnut mare. "Please!" she cried, but they could not hear her. One of them said to another, "Listen! They're sending our corn out to Switzerland, while we eat this and starve," and the other looked at the mare, whose remaining eye let out one last tear.

And then the whisper grew and grew, and the sound of the whispers was like a million million hailstones falling onto cobblestone streets. The whispers turned to shouts and screams and then there were the sounds of a rushing river and a million million windows smashed.

The youngest daughter cried out again, and when the mob had left, she took the last remaining piece of horse. She climbed up the house with the piece in her mouth.

She leapt for days and when she came home she slid down the chimney and cried out, "Here, here you go, Mother, eldest daughter," and as she did so, the mare fell from her mouth. Her mother and sister scrambled to eat it from the floor. A small piece was left for the youngest daughter.

A day passed, but the horse was not enough to still their hunger. So the mother said to them both, "*You've got to die or else I'll waste away.*"

"*Dear Mother,*" the two daughters said, "*we'll lie down and sleep, and we won't get up again until the Judgment Day arrives.*"

And *so they lay down and fell into a deep sleep*, from which they could not—or would not—awake. The mother left, amidst the rain and the hail and the rioters and the rich clamouring on their hill. No one knows where she went. Some say she joined the sect that sailed down the Danube as far as Ismail in the Ukraine, where they remained stranded, as their food ran out. Others say, no, no, that took place earlier, she could not have gone until the summer of 1817, she was part of the group that made it all the way to Russia or to Brazil. And still others say, no, she was one of the thousand that died in just one day of disease. And still more say no, she landed in a

Dutch seaport, but with no funds to buy her passage, she was turned away. She begged and stole her way back across the Rhineland, until, tired, she curled up under a tree and sighing, died. Some say that you can see this tree there still. It is a gnarled poplar. Right underneath it, a purple thistle blooms.

Let Not the Lord

I

Here's gorse and sunshine and wind; ruins of stone crofts, villagers long gone from this forbidding cliff. Their old homes riddled with sheep, and so here they were, hacking at the dirt and making new homes from the stones unearthed. There was just one horse. Men pulled the harrow. Harrow: a heavy frame with tines, to pull over the ploughed field, to break up clods, to remove weeds, to cover seeds. From Old Norse, "herfi." Related to Dutch "hark," rake. Archaic: pillage, plunder. To disturb the mind. To cause distress.

Hark to this earth, this man and this wife spinning, or hauling manure, or gutting fish. Ropes tether the children to cliffs, so they can't blow off. This son of mine. We're on this cliff and he's tied to my chest. He's solid, and yet the fear, the dandelion fuzz. We want to hold on.

You think sometimes that things are holding still, or that just one thing is happening. That the volcano is erupting. That the Thames is freezing. That these men are fishing. That this couple here is drinking coffee, and all that is happening is the coffee in the cups, and their conversation, but all this time, the earth is changing, the babies and men and women are blowing off the cliff, or being held on by love and rope and goddamn determination.

You can almost hear, above the sounds of their play, the rope creaking, the strands thinning out.

In 1814, there are 80 inhabitants, and Mary Wollstonecraft Godwin leaves Scotland and goes back to London, where she meets Percy and Harriet, and in November, Percy and Harriet's son is born. Two months later, Byron and Annabella marry, and a month after that, Mary and Percy's first baby lives, then dies. Mary stays in while Percy and her stepsister sail paper boats down the Serpentine. It is February and the lake is clear.

December 1815 and there is Byron's daughter Ada squalling, and there are squalls outside, so no one is outside at all. There is the stink of peat, and the children are wearing woollen sweaters. The village's name sounds like Bad Bay. The waves crash upon the shore.

Dream that my little baby came to life again; that it had only been cold, and that we rubbed it before the fire, and it lived.

And even if the child is tossed off the cliff and transformed into a partridge, this is no consolation, for we wanted her to stay herself, even as the arms were turning to wings and the eyes were beading black. What we wanted was you, awake, and warmed. We wanted not to be birds, but to be ourselves, enhanced, winged.

*

Ada, at 12, to her mother. Her father never known and long dead.

I wish that supposing I fly well by the time you come back you would, if you are satisfied with my performance, present me with a crown of laurels, but it must only be on condition that I fly *tolerably well.*

Since last night I have been thinking more about the flying, & I can find no difficulty in the motion or distension of the wings, I have already thought of a way of fixing them on to the shoulders and I think that they might perhaps be made of oil silk and if that does not answer I must try what I can do with feathers.

I have now decided upon making much smaller wings than I before intended, and they will be perfectly well proportioned in every respect, exactly on the same plan and of the same shape as a bird's. Though they will not be nearly enough to try and fly with, yet they will be quite enough so, to enable one to explain perfectly to any one my project for flying, and will serve as a model for my future real wings.

*

Icarus, my heart is aching.

*

This, too: that I can find no record of the name of that first, dead child.

*

About two hours' drive south of Badbea is the town that David Thomson writes about in his memoir, *Nairn in Darkness and Light*.

The sand governed the water. It changed the course of the rivers Nairn and Findhorn, swamping farms and leaving the harbours dry. In October 1694 came the great sandstorm. It was as a river. Men reaping barley ran from the fields and within a few hours the barley was smothered. A ploughman, *almost suffocated by sand,* deserted his plough halfway up to a furrow, pulling his horse away. People ran to their houses and in the morning

almost everyone woke in darkness. All doors and windows were blocked with sand. They had to break gaps in the back walls of their houses to get out… there was no grass on Culbin now, no greenery of any kind, no yellow corn, only miles of sand… on the marsh… lay the corpses of hens, rabbits, hares and even sheep and when the wind dropped there was a pestilential stench of rotting flesh and vegetation.

During the lull in the storm, people drove their animals inland, to look for grass. They then went back to their houses to rescue their things, but the wind picked up and the sandhills shifted. The gale lifted the sandhills and blew them along into *massive, blinding clouds.* The world was ending. They ran for their lives.

The next day, you could not see even the tops of trees. For 250 years, the sands kept moving, shifting, revealing here a dead tree, there a living fruit tree, blossoming till it too was buried forever. In 1798, at the same time the Clearances farther north were forcing people to move to Badbea and work on the cliffs, the laird's house reappeared, the main chimney just visible. A man climbed up and shouted down. From the depths of the ruin, a voice answered and he ran.

52 years later, farther north, in Orkney, an entire Neolithic village was discovered during such a storm, the wind

uncovering beds, dressers and cooking utensils made of rock, 5,000-year-old middens. I held my son there too, strapped to my chest. The sun was shining, and the wind was blowing against us.

*

Alexander Brodie of Brodie, Laird of the lands adjoining the Barony of Culbin, Scotland:

1655 5 Julie, This day was a verie great floud, and delug of rain, which raisd all the waters to a great height. Let not the Lord destroy a land and a people that ar dround in sin and ingodlines.

*

On July 21st, 1816, there were hailstones in Lancashire that shattered windows and destroyed most of the vegetation in the area. On that same day, there was almost total darkness in Argyllshire, Scotland. People got on their knees to pray.

That same month, the Queen of Sweden presided over a march of 6,000 peasants to the cathedral at Bex to pray for deliverance from His wrath. And in Paris, 80 young women glided through the streets holding lit candles and praying to the patron saint of the city.

Let not

Let not

Let not the Lord

"Eli, Eli, lama sabachthani?"

The flames waver.

*

And in Switzerland that winter, the snowfall levels on the mountains were so high that the twin glaciers around Chamonix, Bossons and Mer de Glace, advanced one hundred feet more than usual. *With slow but irresistible power, the ice pushes forward vast heaps of stones, bends down large trees to the earth, and gradually passes over them.* In a few years' time, the Bossons glacier had submerged five hectares of farmland. Residents of the village of Monquart placed a large cross at its end.

Everything is shifting, the sands, the snow, the ice: it keeps changing direction, changing rivers, moving forward or sideways, and we're holding on to this tree and horse and cross. The children are tied to the cliff. Their voices are lost in the wind, and soon enough we'll all be gone from here, leaving just a plaque and a handful of grey rocky walls enlivened by bursts of gorse.

II

There is a legend, told to Claire Clairmont and Mary and Percy, of a priest who took his mistress across the mountains, where God punished them with an avalanche. On some nights, you can still hear their moans drifting across the waters.

They were in a closed carriage, with four horses pushing their bodies through the snow, and ten men digging them out of drifts. The snow pelted the windows. It was May. *Never was a scene more awfully desolate,* wrote Mary. *The trees… are incredibly large, and stand in scattered clumps over the white wilderness; the vast expanse of snow was chequered only by these gigantic pines, and the poles that marked our road; no river or rock-encircled lawn relieved the eye, by adding the picturesque to the sublime. The natural silence of that uninhabited desert contrasted strangely with the voices of the men who conducted us, who, with animated tones and gestures called to one another in a patois composed of French and Italian, creating disturbance, where but for them, there was none.*

*

A thunderstorm on Mont Blanc, and the monster scaling the mountain. In the flash of light, Frankenstein sees him hanging off the rocks.

Lake Geneva like *a vast sheet of fire.*

It is a book that goes places. To Germany, where the monster is created, to Switzerland, where he kills William and later Elizabeth, to a far island of Orkney, where a bride is partly created, then destroyed, to the Arctic, where the tale begins, and ends. In the valley of Chamonix, Frankenstein sees that *immense glaciers approached the road; I heard the rumbling thunder of the falling avalanche and marked the smoke of its passage.* In the morning, he wanders through the valley and sees that the glacier with its *slow pace is advancing down from the summit of the hills, to barricade the valley.* It is here amidst the ice where the monster starts his tale.

*

When you read about that summer in Switzerland and the ghost story competition, there is no word about where baby William is. This is something that would not have bothered me before becoming a mother, but now I want to know, always, where are the children? Especially the ones who won't be alive much longer. The adults are by the fire at Byron's villa, Diodati, and there are women who take trips in sailing boats with telescopes to see if they can find ladies' stockings drying or other evidence of female habitation, and I too want evidence: diapers, shifts soaked with breastmilk, a whole room that smells of milk, a basket of swaddling blankets, pins. I want to know, when Victor Frankenstein is sneaking down to charnel houses, where the baby is. Where is William when Shelley runs out of the room, shrieking because he had a vision of Mary's breasts with eyes instead of nipples? Did William suck on those breasts, is the nursemaid a wet-nurse,

where is he, where is this baby—

Is he sitting up? Is he starting to crawl? He must be sleeping. Does he sleep well? He's six months.

*

There are troops of wolves among these mountains, wrote Shelley. *In the winter they descend into the valleys which the snow occupies during six months of the year, & devour everything they find outside of the doors. A wolf is more powerful than the fiercest & strongest dog.*

I dreamt I was holding hands with a black bear and I was walking far away from it when it tugged on my hand, as you would for a puppy, or a dance partner, or a child who was getting too far away. He tugged me close. I stood under his jaw.

I didn't know if he'd tugged because he wanted us to walk together, or if he was about to eat me.

III

That August in Scotland there was an earthquake, *which threw down the chimney pots of many houses, twisted the old steeple, and set the bells a-ringing.*

Beds heaved and rocked; and the noise… was like a heavy weight sliding down house-roof. The night was *hazy and calm.* In Inverness, a man sat in a chair and felt it move back and forth, *as if some Herculean person had taken it up with both hands from behind, and shaken it violently.*

In the mountains south of Relugas, there was a *noise as of a rushing wind, which came sweeping up the hills like a roar of water.* In Perth, there was *a forward and then a backward motion of the earth.*

Church bells and *bed curtains moved as if by wind.*

Dogs howled and poultry on the roost manifested much dismay.

The whole summer had been… wet and stormy. The rains fell, and kept falling, and yet on this evening, it was *fine and still,* though some saw a flash of lightning after the quake, and at Dunkeld, they saw *a small meteor* pass from east to west.

Mr Gilfillan of Comrie states that there was an uncommon phenomenon in the air, —a large luminous body, bent like a crescent, which stretched itself over the heavens.

The next day, *a thick rain came on,* and didn't stop for over sixty hours. *For the next month, there was hardly any fair weather.*

Such a rainy season… has hardly been remembered by anyone.

*

In Ingostadt, Frankenstein runs from his creation, out into the dismal streets. *The rain… poured from a black and comfortless sky.*

*

Wolves descend and all flowers are submerged. 1816 is the wettest Swiss summer on record, with 130 days of rain out of 152. Lake Geneva floods and in the town of Vevey, the *Lower part… was entirely inundated the Lake having risen with uncommon fury to an unusual height – Many Houses washed down & Trees torn up by the roots, the poor people running about in confusion wringing their hands & crying.*

Dead animals float downstream.

The ground is so soaked that fountains bubble from the earth and new streams form. Gardens are under water. Women try to save their vegetables. Men bring hay home in boats.

There is widespread famine and because the country's land-locked, it takes much longer for food to arrive. The government prints out pamphlets advising people what grass and weeds are safe to eat. Thousands of peasants wander the country and beg. Sir Stamford Raffles, having left Indonesia and now travelling in Europe with his brother, writes, …*The great increase of beggars… chiefly children…* [was] *truly astonishing.*

In the district of Glaris, a priest remarks, *It is terrifying to see these walking skeletons devour the most repulsive foods with such avidity: the corpses of livestock, stinking nettles, and to watch them fight with animals over scraps.*

And who promised us a comforting sky? What does it mean, a sky that comforts? Uranus, his eyes just blank stars of dead light.

*

My dearest Hogg my baby is dead… It was perfectly well when I went to bed—I awoke in the night to give it suck it appeared to be sleeping *so quietly that I would not awake it—it was dead then, but we did not find* that *out till morning… – Will you come—you are so calm a creature & Shelley is afraid of a fever from the milk—for I am no longer a mother now.*

And then, ten months later, Mary's pregnant with William. A biographer writes that she was in the early stages of pregnancy later, when she and Percy married, on December 31st, 1816. This was *her third pregnancy in two and a half years.* The nameless one, Willmouse, Clara—all gone by 1819. When William died, she was pregnant with her fourth.

Rock through window. Starving children abandoned, and others killed, by parents. Prisons full. Burglary, embezzlement, arson. Men whipped. Executions, beheadings: man who sets a barn on fire. Another, thief.

In Switzerland, 1817, there are more deaths recorded than births. You can see the faces in the firelight as another shop burns.

*

And ice coats the windows, so you look outside and see nothing but ice.

*

Here's Ada Lovelace as a child, with her arms up above her head, and wings made of feather and silk. Perhaps a mask, too, a masquerade one—light blue and purple, with a long and narrow beak.

Hay washes down the stream. "I am the boss of the glories of nature!" my son cried, and a whole crop gone. All the grapes failed, and fields of wheat and potato under water, and no fodder for the cows.

Her feathers are iridescent, like the glitter of sun on sand, or on snow.

IV

Five hundred years before the year of no summer, the four-century-long Medieval Warming Period ended. The winters became so cold that pack ice extended from Greenland to Iceland. Polar bears walked from one to the other.

In Northern Europe, the rain started in April or May of 1315 and didn't stop until August. It rained for one hundred and fifty-five days in a row. People, cattle, mills, bridges, villages, and the Suffolk port of Dunwich, where hundreds of houses are reclaimed by the sea.

So beautiful and strange. And surge, storm surge, the great rains. For two years, it rained. Quarries too wet to be mined. Wood and peat too wet to burn. Meadows too wet to be mowed. Sod too wet to be cut. Crops too wet to be planted. *The floods of rain have rotten almost all the seed*, commented one contemporary. *In many places the hay lay so long under water that it could neither be mown or gathered.*

Ten inches of topsoil wash away from Europe, leaving sandy clay or bare rock. Half of all arable land disappears.

There is not enough sun in the sky to dry the tide to salt. The cheese and herring trades suffer, the fish floundering at the bottom of the boat, and the rain making waterfalls from your hat. You are behind a curtain.

You are behind a curtain and there's nothing you can do about it, but walk barefoot in this religious procession. For so it has been ordered. The men are naked. Priests lead. You carry bodies of saints and relics. You show your contrition before God.

Land into sea, and marshland converted to cropland converted to marsh.

What bird is it, then, that comes back with soil in its beak?

No, not soil, but an olive leaf.

Nothing that can or should be eaten. Battles rage and there are thefts: livestock, grain, ale. Looting, extortion. Fields full of hallucinating men eating wet rye.

The poor *gnaw like dogs on the raw dead bodies of cattle and graze like cows on the growing grasses of the fields.* And stories of mothers, eating their young.

*

A dove with an olive leaf in its beak.

"Dovecote" is a beautiful word. There are holes for the pigeons or doves to nest.

And the coo and gabbling sound lovely, I hear it sometimes in the bell tower at the park, when the bells aren't ringing to commemorate the dead from another war.

Only nobles were allowed a dovecote. The rest were forbidden by law.

The papers are full of complaints by peasant farmers who said that the doves stole seeds during planting time.

The ark is behind a curtain, and the bird is flying towards it. This is a new world. Where even a dove is political.

*

"Flanders" is an old Dutch word for flood. And in 1315, in Flanders, the rain starts in May and doesn't stop for more than a day at a time. The mud stops the army. Horses drown every day. Food rots. In Ypres, in thirty weeks the number of dead from the famine is almost three thousand. In just one week in August, there are 191 gone.

How can you tie children to cliffs so they don't blow off and then later eat them? Or is that earlier?

Time rushes down the gutter, falls from your hat. Oh my child, you were just cold.

The Ice Maiden

There is a way of reading the earth. Every tree and stone, and deep in this dark ice. The ice crept forward, little by little, every year a bit closer, and the men and women also, creeping towards the creeping ice. There was a minister, he stood there with his book held in his right hand, like so, over his heart: his chest and the thick book.

But the thick ice said nothing.

Oh, but that is not so, for it spoke, it crackled and cackled and the women underneath the ice, and the men too, waved their thin fingers of ice, and it was like a beckoning: "Come, come to us." Or maybe it meant "go away, back off, the ice is the strong one here." A sign can mean both one thing and its opposite.

But the prayers were so loud, no one heard the ice or saw the drowned men and women, with the snow in their eyes and the ice in their throats. Poor Rudy and Kay, the one drowned in the blue-green glacial water, the other with a speck of glass in his heart. Everything was sparkling cold.

The prayers looped around and around like the rings inside a tree. You can read those rings, when the tree is cut open and the heart of the tree is broken. The sap drips down: look, it is bleeding!

There is a story of an artist who cut the thinnest slice of tree, like the thinnest and tenderest slice of roast, and he placed it on a turntable. He made the record of the tree into a record: that is to say, he took the language of the tree, the years and the seasons, and he translated it into piano music, so that the rings could be heard. The widest, darkest rings are the good years, the narrowest and lightest the bad, when the harvests failed and the people starved and the glacier crept forwards like a cat, twitching his whiskers of ice.

The snow was fierce in every corner, and the wind blew so bitter and strange, and the people got smaller and smaller, like so.

And the tree ring was so faint, it wasn't there at all. When the artist tried to play that year, it just skipped on to the next, proving, remarks Andrea Pelster, that *even a tree will only tell the story it wants to tell.*

The procession came here two hundred years ago, with their thick boots and thick shawls and book, and the man came and blessed this glacier. Then this Year of Our Lord, where history loops around and around, with this cross planted at the end. The man prays to the gods, or God—

but the ice moves forward, and the inhabitants look on. There are several dwellings under it now.

It could be not to this God, but to an earlier one, a god of ice, or to the Ice Maiden herself. "Vanity!" she grumbled from her palace, "Vanity!" She laughed, and every time she laughed, the people in the valleys far below said, "There's an avalanche. Why, there's another one."

In the water, writes Hans Christian Andersen, you can hear *the echo of church bells from buried towns.*

"They're pretending to be gods down there, those powers of reason," said the Ice Maiden. "But the forces of nature are still in control."

And here came a learned man who said, Yes, *this earth which we inhabit will at some future period be changed into a mass of frost.*

Oh, you can scream and cry and make these funny gestures on your chest, or haul this cross over your shoulder to plant at the end, but the ice will advance, it will knock over your cross and your learned men and your credulous fools.

There is a dam made of ice, and beyond that is a lake, and should this weather warm, the lake will get larger and larger till it bursts free of its bonds.

Why even as we speak, the lake is getting larger—two miles long, 650 feet wide, 200 feet deep—and the water is even now pushing against its prison of ice. If it bursts, the flood—twenty million cubic feet of water—would drown the town twelve miles away. It will destroy farmlands, villages. ("Vanity," she laughs, "Vanity!")

They sent a man while the snow swirled and the branches were heaped with frost. Under his direction, the workers cut a tunnel through the wall of ice so the water could slowly seep out—and so, the men, soaked in icy water, with blocks of ice as big as another man, raining down upon them from the glacier overhead.

The men stood in the water that was ice, and their sharp hatchets were dulled by the ice. Their shoes were soaked, and then it snowed, two feet in just two days, and this in May, and so, the slow procession of many out again, into the dry world.

In June the water began to come through the tunnel. Three days later, it had dropped by 30 feet, and by the time the dam burst, the tunnel had made the volume of water drop by a third.

In the afternoon of June 16th, 1818, the dam burst. Twenty million cubic yards of ice and mud came hurtling down. It came roaring into the valley, dragging behind it rocks, great blocks of ice and tree trunks, their roots like giant worms scrambling for the dark. Entire forests were ripped up by the roots. There was a thick black fog. Roads, bridges, orchards, sawmills, houses, furniture, flour mills, ironworks sank below the flood. The dead lay scattered.

It took half an hour to pass. It spilled across the river Rhone and was absorbed into Lake Geneva by midnight. What before was *so beautiful and so populous was converted in a moment into a dreary desert.*

Oh, the thick black fog, and beyond, the tintinnabulation of vanished towns.

You can find the temperature of the world by reading the rings of trees and also the core samples taken from ice sheets and glaciers. There is a story written there, of annual snowfall and pollution. Toxins get embedded in the ice, and so, deep in the Greenland ice sheet, you can read about Chernobyl and atomic bomb testing and the use of lead in Roman foundries. It is a layered rock.

Through it, you can tell that there was another eruption shortly before Tambora, the one that scientists dub "The Great Unknown." This occurred in 1809, somewhere in the tropics. It was the third-largest since the early 1400s.

But that is another tale. For now, *there is nothing, the Ice says, but Ice.*

The Laufmashine

I

Here's a story. In the summer of the Hunger Year, Baron Karl von Drais unveiled his Laufmashine, a.k.a. the running machine, draisine, velocipede. It was made of wood and two iron wheels. There was a single cushioned seat and a padded board on which the rider's elbows could rest. This was the precursor to the bicycle: instead of pedalling, the rider ran. In good conditions, he said, you could run up to twelve miles an hour, as fast as a galloping horse.

This, so far, is true.

At a demonstration in Mannheim, Drais covered nine miles in an hour. A month later, he ascended a steep hill between Gernsbach and Baden.

In his pamphlet, Drais advertised different models. There was a tandem one and *a three-wheeler with a second seat in front*... [for]... *a lady passenger*. There were optional accessory items, like umbrellas and sails for windy days.

A critic scoffed that the Laufmashine turned men into horses and carriages, but I also picture boats. If the wind picked up, you could, if you were the passenger, see it billowing through the sails and imagine that this man running was a ship. You are sitting up on deck, watching the dirt, the mud, the sea below. On this road, the crow's nest. In this vast sea, a tree.

Two years later, Gaetana Brianza of Milan designed the Velocimano tricycle, which had the head and neck of a horse, and the wings of what looks to be a dragon. The rider operates a pair of levers to work the drive connected to the rear axle and to the wings. The wings flap as he operates the drive.

In the picture, the man holds the wings by means of loops, like reins. The horse has a slightly open mouth, and you can see his teeth. He looks like he's smiling.

I have made the horse male. I have made the wings reptilian. With their strong lines and scalloped edges, they look nothing like a bird's.

It's possible of course that these wings are exactly the sort a horse *should* have, were he to fly. Certainly they look heavy, and you would need something muscular, veined, and tough to the touch to lift such a large animal up to the skies. The wings look like the spines of an umbrella, covered by tarp. They look hard and real and exactly the opposite of flights of fancy, nothing so gossamer as that. If you pull on the reins too hard, if you squirm in your seat, that horse will turn his head and breathe out fire. Your body will be ash.

Mostly they look like the kind of wings that could get a horse out of the mud, when his mouth is wide open and the guns are going off. These wings will never melt.

*

In 1818, at another exhibition of the Draisine, bands of children had no problem keeping up with the rider. One critic remarked that this invention was no more practical than the mechanical wings exhibited by another the year before. Quipped a second: *Mr. Drais deserves the gratitude of cobblers, for he has found an optimal way to wear out shoes.*

That same year, Drais rode more than fifty miles from Mannheim to Frankfurt, but by then the press had moved on.

*

Here's where we veer, perhaps, from the truth—but what is truth?

1815: a blast, fireworks, pirates, and the people and the horses died, as they did the next year in Germany, and the next, and the next. There were no oats to feed the horses, and the people and the horses died, the ribs showing through thin, un-tarp-like skin. The skin was tissue paper.

Those horses that didn't starve to death were eaten, and because of this, when corn bought in Russia or the Netherlands was shipped to the Rhine Harbour in Mannheim, there was no way of bringing it to the interior, because, says one writer, *There were no horses left.*

The hollowness of that. It's Mary Lennox in the bungalow in India, with nothing alive but her and a jewel-eyed snake.

We have no statement, the account goes on, quickly, as if we aren't picturing the corn lying unused in the harbour, or the workers on the docks, opening burlap sacks and running their fingers through it like misers with gold, while in the interior, the family lies in the one room, their eyes glassy, their stomachs distended, and in the fields, the crops lie in ruin—and no, no horses run along the meadows or in the flooded roads, because *there were no horses left—we have no statement*, because the narrative must go on, and this is of course about the history of the bicycle, not of a family that has, why am I picturing, a cat? Or maybe it's that the baby just sounds like one? Mewling and mewling away?

We have no statement—but let's make it a cat, thin-ribbed and yowling, bursting from, no, not the house, but the half-gone barn—it was a farm cat, of course, orange with a kink in its tail.

Thus, there was a need for a horseless means of transport.

We have no statement from Drais on this, but we have circumstantial evidence from his inventions after conspicuous crop failures (see figure 1.4) and from newspaper voices of the period.

There were no horses left.

*

In 1813, Drais built a four-wheeled vehicle that allowed two to four passengers. One driver operated the steering by means of a tiller, while another worked the axle with his legs and feet. This invention, says another, was in reaction to the crop failures of 1812 and the subsequent high price of oats. Because let's not fall into that way of thinking that crop failures started in 1816, 1817, or that barons invented horseless means of transport only after a volcano, or only because of its resulting weather pattern, the brown snow or ash upon our open mouths, a specific set of circumstance.

The next year, 1814, he showed the vehicle to delegates at the Congress of Vienna, convened there to partition Europe after the downfall of Napoleon. His fellow delegates were not impressed, and Drais, disheartened, went back to Baden and invented other things, such as, eventually, the meat grinder and the first typewriter with a keyboard.

In 1817, he created the two-wheeler. Some say he did this so it could navigate the forest paths, where he briefly worked as a forestry official. Drais himself remarked that *The main idea of the invention* [was] *taken from ice skating.*

Two ladies and a gentleman skate a reel. The world is hushed, and along this forest path, the snow lies heavy on each branch.

Das Hungerjahr, and in the towns, the children run after, laughing and pelting the velocipedic dandies with stones.

II

It was the birds that woke me. I live on a busy street, and throughout the day, drivers use it as a shortcut, gunning up the hill, and so you'd think it'd be trucks that woke me at 5 a.m., but no. In a city, you cannot clearly see the stars, but you can still hear the birds, who will go on singing, regardless, it seems, of the cacophony of car or machine gunfire. That driver's rushing to work, there's a girl dragging her sister up the street and the little one is wailing, "Daddy, Daddy!" and I can't see where her father is. I am in the garden reading a memoir about the Great War, where every man the girl loved—her fiancé, her brother, their close friends—has been killed.

I'm sitting over an anthill and the flies and ants are crawling up my legs and the girl is crying, and motorcycles zoom past and the birds—chickadee? sparrow? thrush?—are singing. But the song changes. A U.S. study shows that sparrows sing louder in urban areas, and their songs have fewer notes in the lower register, notes that cannot be heard amid the clatter of the city.

There's a story of Vauxhall Gardens, where there used to be birds, and people would come and listen to them sing. In 1827, a German Prince wrote of Vauxhall, of its *illumination with thousands of lamps of the most dazzling colours.* On the trees there were lamps of *red, blue, yellow and violet,* made to look like *large bouquets of flowers,* and there was *a temple for the music* [and] *the gardens* [that] *extended with all their variety and... exhibitions, the most remarkable of which was the battle of Waterloo.*

A battle, victory, jingoism amongst the buds. There's nothing here of loss. It's the replica of Trafalgar at the fair again, the crowds that cheer as the little ship sinks, the stalls filled with apples. It's like the time I went to a marzipan museum and saw the Austria-Hungary empire as it was before the war, laid out in sugar, with no reckoning of its effects.

*

In the 17th Century, Vauxhall Gardens was a veritable aviary, where nightingales would sing back to violinists. Later, when they and other birds began to disappear from London, it was said that men got paid to hide in the bushes at Vauxhall and imitate them, so as to contribute to the rustic atmosphere. People came to hear birds, and did.

But if the nightingales were indeed disappearing, it would be not because of noise, but rather loss of trees. After all, it was not just guns that killed off the passenger pigeon, but urban sprawl upon the land. And while guns aimed at birds will end the song, the noise of guns themselves might only change it. The Western Front was full of nightingales, singing amidst the guns. Perhaps like the sparrows of San Francisco, they were singing even louder than before.

*

There is a bird called *Bonaparte's Gull*, so named in honour of French zoologist Charles Lucien Bonaparte, nephew to Napoleon and vital contributor to American ornithology.

Still, when I flip through the pages of this bird book and see this, I think of the war, these different wars.

It is the *only gull that regularly nests in trees*, her nest an *open cup of twigs, small branches, and bark, lined with mosses and lichens.*

I listened to the call of Bonaparte's Gull, but could not explain it. For J. Fenwick Lansdowne, *the sound... reminds me always of a reel as a salmon takes the lure and runs with the line. I do not know how else to describe it—a repetitive buzzing whine, heard most frequently when the birds are foraging or resting on the shore.*

When these gulls are *disturbed at the nest, their screams are loud and prolonged.*

*

Bonaparte's Gull was first described by George Ord in 1815, the year of Waterloo and Tambora.

A year earlier, Drais showed his four-wheeler to men at Congress, meeting to divvy up the spoils of Europe:

is this the time to show your newfangled invention?

That's before Napoleon escaped from Elba, before Waterloo. On the eve of Waterloo, wrote Napoleon, *the rain was falling in torrents. On the roadway the soldiers were halfway up to their knees in water. On the surrounding ground they sank up to their knees. The artillery could not get through and the cavalry could only get along with difficulty.* Mud was made worse by horses and men whose formations, from square to column to square, changed over and over on the same spot. Men lost their boots in the mud and could not stop in their formations, so marched, bootless:

Delay the start to battle. The ground is too wet to maneuver the guns easily.

Still, the battle occurs and the mud remains, worse as it continues, so that the Old Guard cavalry, launched near the end, slog through a thousand yards of mud, easy targets.

*

Is there anyone who wouldn't scream, loud and prolonged, when disturbed at their nest?

*

Nothing quite so neat as this. But still:

The weather turns, the crops fail, the price of oats rises, the horses cost too much to feed, the inventor creates a horseless means of transport, the men laugh, Napoleon escapes, the army forms, the volcano erupts, birds, horses, men, the weather turns, the mud grows, the boots stick, the summer never comes, the crops fail, the price of oats rises, the horses cost too much to feed, and are either eaten or starve, the inventor creates another horseless means of transport, the men laugh—

She stood out there during the armistice celebrations, and thought that *in this brightly lit, alien world I should have no part. All those of whom I had really been intimate were gone; not one remained to share with me the heights and the depths of my memories. As the years went by and youth departed and remembrance grew dim, a deeper and ever deeper darkness would cover the young men who were once my contemporaries... The War was over; a new age was beginning; but the dead were dead and would never return.*

III

At Waterloo, the sound of the guns in the distance was likened to the rolling of the sea. From afar, you could hear the horses screaming—

and screaming, when, a hundred years later, 6,500 horses and mules died in the rolling of the sea, or before their bodies flew into the air and hit the water, in transports attacked by submarine.

It's always an even number like that, neatly rounded off.

And half of those who made it to the Front were galloped right into machine guns or died while carting guns and food and wounded men—

when the shells burst, there was no place for them. They stood in the open, their tails collecting ice.

Mud and slush, the *white light of bursting shells… Above all, there was the smell of blood, terrifying to every horse.*

Twelve in one heap. And the mule's forelegs were shot away. The man watched, trapped, and longed for a shell to finish it off. It was stuck in the mud, *trying desperately to get to its feet that weren't there.*

*

There was a gas attack and *nearly every green thing* withered and the horses and mules had *water running from their mouths and eyes.*

The rain came down, and the men were smeared with chalk, their faces white with it. The rain came down, and there was gristle caught on bushes. The rain came down, and in our way was a dead horse with *smoking entrails*. The rain, and the trench falls down in places, and you see a mass of frogs and slugs and men, dead from months or years before. The dead are stacked in layers, and here they will rest, till the bombs come back again.

We are up to our waists in ooze… On the right flank a dead body is coming to light, the legs only so far.

Entire trees uprooted and tossed into the air, a whole forest on fire. Fountains of mud blown high up and falling back in place: this mud volcano. Over the torn earth, a yellow and brown creep of gas and mist.

It was all a ghost story, writes Edmund Blunden, and I think about the will-o-the-wisp, that *haunt burying places, places of execution, and dunghills.*

Some that have been catched consist of a shining viscous matter, like the spawn of frogs, not hot, but only shining.

*

Every dawn, something new, a different wilderness.

A white tomcat with his paw shot off visits no man's land. A carrier pigeon, frightened of gunfire, flies back into the dug-out. The men find him huddled under the floor.

*

The horses were hungry. Sixty percent of the Germans' trucks broke down, and so they used horses, but the horses needed two million pounds of feed a day, much more than the French countryside could provide. The horses ate green corn from the fields and died in the tens of thousands. And when they died, there was no one to bring the supply wagons and the men started running low on ammunition and food. This was at the beginning, when they were on the move: before the war became, in Siegfried Sassoon's words, *a matter of holes and ditches.*

Back home, a blockade meant that twenty-five percent of food was no longer available. Crops were stunted because half of the fertilizer used was also imported. At the end of 1916, there was a bad spell and nearly half of the potato crop was ruined. They called it "The turnip winter." *More than 50 food riots erupted*, writes Adam Hochschild, and yes, "eruption" is the word here, for all of this.

When a horse collapsed in the middle of the street in Berlin, *women rushed towards the cadaver as if they had been poised for this moment, knives in their hands. Everyone was shouting, fighting for the best pieces. Blood spattered their faces and their clothes… When nothing more was left of the horse beyond a bare skeleton, the people vanished, carefully guarding their pieces of bloody meat tight against their chests.*

*

Sometimes the sun shines and flowers bloom. And in every weather are the pigeons, fluttering and alighting onto the garden tomb. There is the robin in the hawthorn, and *in trampled gardens the yellow rose, the blue-grey crocus.* The blackberries are ripe *in the low hedges.*

Tree trunks are spikes and daily, apples are gouged out by shells.

My son, my son, wipe your hands against this clean cloth, the mud and blood off of your face.

IV

Blunden rode his bicycle into the Somme. *The heavy machine went slower and slower, and stopped dead.* He was thrown off. The brake clogged with mud.

And at the German intelligence station, there were despatch-riders, bicyclists, carrier pigeons. Four bicycles that were left outside got twisted, broken, scattered by shell.

The yellow and chrome bicycle by the small-kneed child, the horses tossing their heads in the barn and along the green road. We're here, we want the sunshine along the path, the red poppies, the boy upon the strand. Years ago, I sat in class and learned about the bicycle.

Horses, wrote a reporter in 1915, *are slow and quickly become frightened and exhausted. It is a common sight to see a patrol pedaling along on low gear on a heavy, muddy road, making good time in comparison to what a horse could do.*

There were men with bicycles folded onto their backs. There were horses with men folded onto their backs. There were broken things and breathing things and men.

Let man have dominion over the fish of the sea, and over the fowl of the air, and over the cattle, and over all the earth, and over every creeping thing that creepeth upon the earth. Let him have dominion over eight million horses, plus mules, camels, bullocks, *many hundreds of dogs, carrier pigeons and other creatures* [that died] *on the various fronts during the Great War.*

*

The horses… screaming with broken legs… or their entrails hanging out… we… spent the next hour… shooting them through the head. To do this we had to wade ankle deep through blood and guts. That night we lost over 100…

But the machine couldn't die. It broke and was fixed or replaced. The men dug ditches and buried cables, pedalling through and through the long night.

V

In 1816, there was a British expedition led by Captain James K. Tuckey of the Royal Navy, to find the origin of the Congo. There were two ships: carpenters, blacksmiths, surgeon, gardener. Tuckey counted eight slavers at the mouth of the river.

They sailed their own ships up and then, to avoid the rapids, went overland. Scaled *almost perpendicular hills,* scrambled *over great masses of quartz.* Crystal Mountains, yellow fever. Twenty-one out of fifty-four gone, and Tuckey himself.

The year of no summer ends, but this continues with more explorers, more men and now there's quinine, steamboats, weapons, a King who in later years rides his large tricycle around his grounds. "*Mon Animal,*" he calls it. "*Mon animal.*"

At first, it was ivory, and there were raids and cheating, people forced to sell or deliver only to Leopold's agents. The porters carried everything. They looked like *skeletons dried up like mummies, their skin worn out... seamed with deep scars, covered with suppurating wounds... no matter, they were all up for the job.*

When you fall off the bridge, you pull the whole chain down, to drown. When you fight, they bring Captain Rom twenty-one heads, to decorate his flowerbed.

*

In 1890, the pneumatic rubber tire is invented, and the bicycle craze begins. There's a worldwide rubber boom. The wild rubber vines twist high into the trees of almost half the Belgian Congo. To get it is hard, painful: climbing high and then drying the sap by spreading it on your arms and chest, ripping out hair. If you do not do this, they attack, burn villages, seize wives and children as hostage. Naturally, they kill. *I always have to cut off the right hands of those we kill in order to show the State how many we have killed... To gather rubber in the district, one must cut off hands, noses and ears.*

For miles, you can see the line of smoke.

The country a desert, no natives left.

The hands come to them in baskets. *When I crossed the stream I saw some dead bodies hanging down from the branches in the water. As I turned away… one of the native corporals who was following us down said, "Oh that is nothing, a few days ago I returned from a fight, and I brought the white man 160 hands and they were thrown into the river."*

An ibis, with black body and white wings, flies over us.

*

You walk more than twenty miles to the European agents, who weigh the rubber. You are paid with a piece of cloth, a handful of beads, a few spoonfuls of salt. You skirt this spot here where René de Permetier has all his bushes and trees cut down around his house, so he can sit on the porch and use passersby as target practice.

By the turn of the century, the Congo produces more than eleven million pounds of rubber a year. In March 1908, Belgium purchases the Congo from Leopold. There are no longer government quotas for hands, but there's still forced labour then, and six years later, when the war breaks out. Disease and famine erupt, continue erupting. Lo, Centralia these many years, this underworld burning.

In Flanders, the new causeway is *swollen with dead mules*. The water below is yellow and brown and strewn *with full-sized eels, bream and jack.* Gas oozes from the crumbled banks.

*

I circle around and around, highlight this lecture with my yellow pen. I came to it early, this knowledge, and then forgot, the notes long gone, the exam done.

Here was a tandem one, a three-wheeler with a second seat in front for a lady passenger. Here's a volcano, here's a war, the mud and the sap of the rubber vine coagulating upon his chest.

The quality of the Somme mud, wrote Edmund Blunden, *began to assert itself.*

When Leopold went to see his mistress, he pedalled on his tricycle. "*Mon animal,*" he murmured, and he patted its handlebars and the horse raised his head and screamed, and his scream held the voices, and the pannier held the hands.

The volcano threw the world into disarray and invention, and the mud threw the man off the bicycle, and the other man said, "Look, look!" and showed everyone the prototype. This was at a time when the men at the Congress of Vienna restored the old regime, and the squares and squares of men trampled the mud down and made it worse, and the sound of the guns was like the roaring of the sea. *The sheer devastation of the battlefield, adumbrating Flanders a century later, struck contemporaries,* writes Paul Johnson: "*The farm of Hougoumont... was totally destroyed, the house and offices burnt and battered with shot, the trees around it... cut to pieces.*"

Blunden, waking from nightmare, his hands maybe shaking, took to the typewriter, a later invention of von Drais', and wrote:

Let us stop this war, and walk along to Beaucourt before the leaves fall. I smell autumn again.

I circled round, and left this here.

Afterword: The Path

Conceptually, both the Morecambe crossing and the Broomway are close to paradox. They are rights of way and as such are inscribed on maps and in law, but they are also swept clean of the trace of passage twice daily by the tide. What do you call a path that is no path? —Robert MacFarlane *The Old Ways: A Journey on Foot*

Behind the glass, there was a stuffed passenger pigeon, skeleton of a dodo bird, and the skull of something I didn't take the time to look at because I was overwhelmed and couldn't focus. *Gone Forever: due to human impacts... by 2100 half of the world's species may be lost.*

Adieu à tout jamais.

In a drawer right below the pigeon, there was a whole tray of songbirds, lying on their sides with the labels affixed to the foot. They were blue and green and unexpected: if you pulled out the tray, there they were, captivity neatly arrayed. There was another tray for plants, a third one for mammals. I didn't stop to read the labels, to see what year that yellow bird was stuffed, to see where the green one came from. *Adieu, bonjour, adieu, bonjour*, the tray slides in, the tray slides out, the birds are here, the birds are dead.

What do you call a path that is no path? What do you call it when something is both here and gone? When the tide erases the path and the path is on the map. The birds that lie here in their muted birdlessness.

What do you call it when I sit here and pull this thing, that, into some sort of order, make it cohere, smoosh it into a kind of wisdom? I tell you, I shout it from the roofs: I am not wise.

This mud that horses drown in, it's like the mud of six hundred years later, it is the same mud but different horses, different men, drown in it. Before the handheld compass, people on the Broomway held stone and thread. They carried a two hundred-foot length of linen thread to get them through the fog, one end tied to a stone, and they unspooled the thread as they walked to the next broom set as marker along the path. If they went the wrong way, they returned to the stone at the last broom, and tried again. It is the labyrinth again, it is Ariadne and Theseus, it is men wandering in the trenches, it is Haig at the centre, it is

I, not knowing much.

All along the birds are singing and all I know is this: when the weather turned or disaster struck, there were always voices explaining why this was so, why the rain was falling now, why the snow was brown, whose side God was on. How can a path be here and not,

how can this war be here,

with that war just over, and it doesn't matter, the men are doing what they always have, looting, pulling out teeth and putting this boot up for sale.

The tide goes in and we've vanished. It goes out, and we've returned. The tray goes in, the tray goes out.

I'm standing here thinking it all fits together, but how or why, I know not. My hands are too small for God.

Notes

Their Useless Wings

13: Title is taken from Lord Byron's poem, "Darkness," written in July 1816 "at Geneva, when there was a celebrated dark day, on which the fowls went to roost at noon, and the candles were lighted as at midnight":

...And others hurried to and fro, and fed
Their funeral piles with fuel, and look'd up
With mad disquietude on the dull sky,
The pall of a past world; and then again
With curses cast them down upon the dust,
And gnash'd their teeth and howl'd: the wild birds shriek'd
And, terrified, did flutter on the ground,
And flap their useless wings...

13: "My God, my God..." – Psalm 22:1-2, *King James Bible.*

14: "The banks of Ohio... undulating and angular... torches of pine-knots... hard gale... no one dared..."—John James Audubon, "The Passenger Pigeon."

15: "Long ago... there was a city where birds died" – Quebec City, after a June 6, 1816 snowstorm (other versions have it as Montreal, while in Henry and Elizabeth Stommel's *Volcano*

Weather: The Story of 1816, the Year Without a Summer it is New York City).

15: "...after angels wavered in the air at Mons" – this myth developed from a short story of Arthur Machen's, published on September 29, 1914, in the *Evening News*. "The Bowmen" was not explicitly described as fiction and is written in a journalistic style. It features the ghosts of English Bowmen from Agincourt coming to the assistance of the British, shooting arrows which killed the Germans without leaving visible wounds. The bowmen were "a long line of shapes, with a shining above them." From this, came the angels which some took as fact. In some accounts, British soldiers claimed they had seen these angels, although subsequent investigations showed that some of these eyewitnesses were not even at Mons.

Machen agreed that others could reprint this story, but when, in 1915, he protested that this was fiction, he was told by clergy to "walk humbly and give thanks for having been made the vessel and channel of this new revelation" (quoted in Graeme Donald's *Loose Cannons 101 Myths, Mishaps, and Misadventures of Military History*). Members of the press also said that to suggest the angels were fictional was tantamount to treason.

16: "There are wilde Pigeons..." – Ralph Hamor, "A True Discourse of the Present State of Virginia (1615)."

16–17: "...they set out to search for pirates" – piracy was very common in many of the Indonesian islands—not just for goods, but also for people. As Gillen D'Arcy Wood comments in his essay, "The Volcano Lover: Climate, Colonialism, and the Slave Trade in Raffles's "History of Java" (1817), "'piracy' served as the basic machinery of the Southeast Asian slave trade." (Essay published in *Journal for Early Modern Cultural Studies*, Vol. 8, No. 2, Climate and Crisis (Fall–Winter 2008), pp. 33–55).

17: "Nothing equal to it..." – Sir Charles Lyell, *Principles of Geology*. Here, Lyell slightly alters another account, that of the Captain of the East India Company Vessel *Benares,* which is quoted in Sir Thomas Stamford Raffles' "Narrative of the Effects of the Eruption from the Tombora Mountain, in the Island of Sumbawa on the 11th and 12th of April 1815."

17: "The young men queuing up…" – Geoff Dyer, *The Missing of the Somme.*

18: "…the suffragettes… toured the country… giving speeches and white feathers…"—Emmeline Pankhurst and her daughter Christabel participated in this practice. Notably, Christabel's sister, Sylvia Pankhurst, opposed the war, which led to her estrangement from her mother and sister. A further discussion of the family's beliefs and activities during WWI can be found in Adam Hochschild's *To End All Wars: A Story of Loyalty and Rebellion, 1914–1918.*

18: "explained [he'd] been discharged…" – Frederick Broome.

18: "That night he cried…": This was Robert Smith. Peter Simkins quotes Smith's daughter, Mrs. J. Upjohn, in *Kitchener's Army: The Raising of the New Armies, 1914–16.* Smith's wife had just recovered from a serious illness, and the couple had two small children with a third on the way. Soon after this incident, Smith joined the army. The pacifist who claimed he'd received enough feathers to make a fan was Fenner Brockway.

18: "as the blood…" – Revelations 16:3, *King James Bible.*

19: "Great frost…" – from Connecticut farmer Calvin Mansfield's journal entry for June 11, 1816.

19: "By the breath of God…" – Job 32:10, *King James Bible.* This was quoted in the July 17, 1816 issue of *The Reporter* (Brattleborough, Vermont). After a synopsis about the events of spring/summer 1816 in North America (frosts, snowfalls, sun spots, etc.), the article then goes on to discuss different theories, before concluding, "Perhaps we can assign no other cause than the fiat of the GREAT FIRST CAUSE and the wisest of philosophers with Elihu, the friend of Jub, 'by the breath of God'…"

Mud

22, 24: "No, no, no life… howl, howl…" – William Shakespeare, *King Lear,* Act 5, Scene 3.

24: "Second Battle of Passchendaele…" – list of photographs here from Canadian War Records Office's official photographer, William Rider-Rider.

25: "Nobody would believe…" and story about bombed-out cemetery – Erich Maria Remarque, *All Quiet on the Western Front.*

26: "The mud was six inches…" – Major Robert Massie, January 17, 1918, speech given at Empire Club of Canada.

26: "Red lips are not so red…" – Wilfred Owen, "Greater Love." This is also quoted in "The Last Frost Fair."

27: "In Flanders, a thunderstorm" – July 18th, 1816, thunderstorm in Ghent. During the storm, the cavalry sounded their trumpets at 9 p.m. to signal their retreat. Although this was usual practice, that evening, people thought it was the 7th Trumpet, a sign of the apocalypse: "Suddenly cries, groans, tears, lamentations, were heard on every side. Three fourths of the inhabitants rushed forth from their houses, and threw themselves on their knees in the streets and public places. It was not without infinite trouble that the cause of this extraordinary terror was discovered." *Times* (London), July 23, 1816.

27: "Glass falling slightly…" – Sir Douglas Haig, diary entry near end of August 1917, from *Private Papers*.

28: "Can God be on… brought in with mud… by the time you get this" – Kate Luard.

28: "And no one in the War Cabinet…" – General John Monash, October 18, 1917 letter to his wife.

29: "You can see the craters now…" – this refers to a Mike St. Maur Sheil photograph of Messines Ridge near Ypres.

Tambora

32: "drawing-down of blinds" – Wilfred Owen, "Anthem for Doomed Youth."

34: "Fillet of a fenny snake…" – William Shakespeare, *Macbeth*, Act 4, Scene 1.

35: "There is a video" – "Rare Journey into World's Deepest Crater, Tambora, Indonesia, Sumbawa" – by RIKSESSION 2012.

36: "troubled confused manner… [pumice size of] walnuts… twenty-six survivors… body of liquid fire" – The Rajah of Sanggar's eyewitness account, recorded in Sir Thomas Stamford Raffles' *History of Java* (1817).

38: "no survivors" – Rik Stoetman and Dan McLerran, "Lost Kingdom of Tambora" *Past Horizons Adventures in Archaeology* (November 9, 2009).

39: "deposits of ash are thin, says the writer" and descriptions of the excavations and survivors selling children for rice, etc. – Emma Johnston, "Up From the Ashes the Lost Kingdom of Tambora," *Popular Archaeology* (June 2012).

41: "five or six... only twenty-six badly burned..." and Komodo dragons – Anthony Tully, "The Great Eruption of Tambora Volcano," *Historypost.com* (2000).

42: "Sir Stamford Raffles prohibited the importation of slaves" – this prohibition took place in Java. In September, 1811, the British briefly took possession of Java from the French (shortly thereafter, the area was possessed once again by the Dutch). During this time, Raffles abolished the slave market in Batavia (now Jakarta) and forbade British officers and company employees from owning slaves. Outside of Java, however, Raffles did little: "his small fleet of gunboats was no match, in speed or number, for the hundreds of prows, with a seaborne force of a hundred thousand armed men, that annually made their way south from the raider strongholds of Sulu and Borneo to comb the eastern islands for slaves and transport them to the slave markets of Jolo and Macassar. Between 1768 and 1848, the Iranun raiders of the Southern Philippines transported several hundred thousand people from coastal villages across the East Indies to these markets." (Wood, "Volcano Lover," pg. 45).

As well, for centuries, during times of famine or other hardship, people who were not captured by pirates for profit sold themselves (or their children) to wealthier homes as a form of indentured servitude (for a set period), or for life, for some measure of stability (food, clothing, shelter). When Raffles shut down the market at Batavia, this meant that those left destitute and starving after Tambora's eruption could not sell themselves to the island. They had lost that particular safety net—and Raffles' ship full of rice (sent four months later) did nothing to ameliorate this.

The Last Frost Fair

46: "photographic collages" – Harry Enchin's "Toronto Moments in Time."

48–51: genteel-looking...stiffened into granite...had an enchanter's wand...Whereas You J. Frost...great violence...equal to the report". as well as many other Frost Fair descriptions are from *Frostiana or a History of the Thames in a Frozen State* "printed and published on the Ice on the River Thames February 5, 1814, by G. Davis."

48: "coral-made" – William Shakespeare, *The Tempest,* Act 1, Scene 2.

49: "If only I could have you back to life" – Brothers Grimm, "Faithful Johannes," translated by Jack Zipes.

49: "His face was smeared" – Jackson Bennet about old pigeoner "Joe."

50: "...the windows frost over so strangely" – Hans Christian Andersen, "The Snow Queen," translated by Tiina Nunnally.

51: "move their unbroken columns..." – Chief Simon Pokagon, from "The Chautauquan," November, 1895, Vol. 22, No. 20.

52: "a loud rushing roar..." – Alexander Wilson, "The Wild Pigeon":

"Happening to go ashore one charming afternoon to purchase some milk at a house that stood near the river, and while talking with the people within doors, I was suddenly struck with astonishment at a loud rushing roar, succeeded by instant darkness; which on the first moment I took for a tornado, about to overwhelm the house and everything around in destruction. The people, observing my surprise, coolly said, "It is only the pigeons;" and on running out, I beheld a flock thirty or forty yards in width sweeping along very low, between the house and the mountain or height that formed the second bank of the river. These continued passing for more than a quarter of an hour, and at length varied their bearing so as to pass over the mountain, behind which they disappeared before the rear came up."

"glittery undulations":

"The leaders of this great body would sometimes gradually vary their course, until it formed a large bend of more than a mile in diameter, those behind tracing the exact route of their predecessors. This would continue sometimes long after both extremities were beyond the reach of sight; so that the whole, with its glittery undulations, marked a space on the face of the heavens resembling the windings of a vast and majestic river."

The Garden of Fugitives

55: This essay describes exhibits from "Pompeii: In the Shadow of the Volcano," seen at the Royal Ontario Museum in October 2015.

55–56: "Ashes were already falling… we had scarcely… You could hear the shrieks… many besought": Pliny the Younger's Second Letter to Tacitus, 6.20, translated by Betty Radice.

Ragnarok

57–58: "Wolf-time, wind-time… a river full of knives… the earth sinks…" – "The Volva's Prophecy" lines 47–50, 36, and 57, *The Poetic Edda*, translated by Jeramy Dodds. The imagery of the sun blackened is also from this prophecy: "Next summer the sun blackens./The weather gets wicked." (lines 43–44).

57: "The Great Winter" – known in Norse Mythology as "Fimbulvetr."

57–58: "bright sun extinguish'd… moonless night… seasonless, herbless…" and imagery of an object (in his case, "the icy earth") swinging "blind" – Lord Byron, "Darkness."

57–58: "black as sackcloth… every mountain and island… I will kill her children…" – Revelations 6:12, 6:14, and 2:23, *King James Bible*.

58: "a ship full of fever" – "ship fever" is a nickname for typhus.

58: "In China..." – in Yunnan, China, in 1815, there were flooding rains, and wheat and barley sprouted under water. Frost in August killed rice fields, and possibly more than 2/3 of the rice crop was destroyed. As a result of the food shortage, prices rose. In Guanyin Valley, villagers ate yellow loam and many died from the resulting "swollen gut." When the "summer" of 1816 arrived, people were already badly off and the snow in July, August frosts, etc., resulted in a complete failure of rice crops. People sold their children on the street. The coldest temperatures in Yunnan were in 1817–18; in 1817, the government dispensed free grain. Regarding opium, one theory is that because farmers couldn't grow rice, they supplemented with the drug—converting their bean and wheat fields to poppy. There was an explosion of poppy farming in the late 1810s.

58: "In Ireland, lice destroys thousands" – there was a huge typhus epidemic in Ireland in 1816 and 1817. The weather was wet and cold, with the Dublin rainfall in July 1816 more than four times the amount of the previous July. The wheat crop failed—the grain was blighted or the husk burst before germination. People sold their clothes and hair for food, but wearing rags made it easier to contract typhus, as lice could circulate more freely. Charles Grant, new Chief Secretary for Ireland: "In the years 1816 and 1817, the state of the weather was so moist and wet, that the lower orders in Ireland were almost deprived of fuel wherewith to dry themselves, and of food whereon to subsist. They were obliged to feed on esculent plants such as mustard-seed, nettles, potato-tops, and potato-stalks—a diet which brought on a debility of body and encouraged the disease more than anything else could have done." In 1817, there were food riots. In Ballina, the army was called in to break up a riot over the export of oatmeal. Three protestors were shot and killed, and many more wounded.

58: "In India, the monsoon" – the sulfate gases from Tambora slowed the Indian monsoon for the next two years. In Bengal, as the monsoon came later, there was first drought, then late flooding. This altered the microbacterial ecology in the Bay of Bengal. According to Gillen D'Arcy Wood, cholera bacterium, "highly sensitive to changes in its aquatic environment" then mutated into a deadlier strain. The cholera epidemic in Bengal in November 1817 resulted in 5,000 men, women, and children dead in a space of five days. This strain of cholera spread over India and eventually, the world.

58: "That God has expressed..." – anonymous editor of *The Panoplist* and *Missionary Magazine* for the Year 1816 in "The Season" (Volume 12). The essay starts off with a retrospective of the weather in the U.S. that year (briefly touching on Europe as well). It ends with: "That the sufferings of all might be mitigated, so far as shall consist with their permanent good, and with the purposes of the Divine government, should be the constant prayer of the Christian. During the cold weather of June and July we had an opportunity of how the minds of people were affected. A general gloom was evident, and a general conviction that men are really dependent upon God. May this conviction become more deep and abiding, and produce its proper influence on the hearts and lives of all."

58: "Hailstorms..." – paragraph describes "Year of Wonders" in 1811. According to one account, the exodus of squirrels was in the tens of thousands. The earthquake measured 7.7 and took place in Arkansas on December 16th (with the Mississippi only seeming to flow backwards).

58: "Church bells ringing in every storm": ministers during this period often rang church bells in lightning storms to ward off evil demons.

59: "The Frost Giants" paragraph – first four sentences' description is closely adapted from *D'Aulaire's Book of Norse Myths*.

59: Beggars "mistaken for an army on the march" – France, 1816. 1817 was known in Germany as the "Year of the Beggar."

59: "Men desired to die…" – changed tense and a few words, but otherwise this is from Revelation 9:6, *King James Bible*.

59: "Great is the distress…" – translation of a German medallion, created in remembrance of the years 1816–17.

59: "Wherever one god falls… moving to the Holy Land… injunction not to bury their dead…" – Spring and Summer of 1816 saw many religious revivals in the U.S. Between 1812–15, there were 500 new members a year in Presbyterian churches in western New York State. In 1816, there were more than 1,000; in 1817, almost 2,000. In Württemberg, Germany, a Protestant sect believed the end of the world was upon them and they needed to emigrate to the Holy Land to escape the plagues of the apocalypse. A band of forty families left in September 1816, sailing down the Danube as far as Ismail in Ukraine, where they remained stranded when their food ran out. (See also reference to this in "The Children of Famine.")

59: The cult where families were urged not to bury their dead was known as "The Pilgrims," which was led by Isaac Bullard who was in turn supposedly guided by visions. Isaac and his family were forced from Lower Canada after an infant was poisoned due to his proposed remedy. In Spring of 1817, they travelled to Vermont, where they gained forty more converts. They then went on through New York, Ohio, and Missouri, gaining more followers along the way. They eventually settled on "Pilgrim Island" on the banks of the Mississippi, where the cult died out, due to fevers and resistance to Bullard's autocracy. The estimated numbers in this cult vary, but some sources say "several hundred" while others argue it was lower. In 1824, there were only two people, both women, who remained at the camp. It is unclear what happened to Bullard.

The Pilgrims rejected material possessions and practised free love. They fasted, ate standing up and were punished by having to stand straight without food or sleep for four days. They were not permitted to eat without Bullard's permission. They also did not bathe as Bullard had not found any mention to do so in the Bible (he himself boasted he had not bathed for seven years). Nor did they bury their dead, because of the Biblical injunction. People in the area objected to the dead bodies on the ground.

59: "Let the dead…" – Luke 9:60, *King James Bible.*

59: "The great day" – Revelation 6:17, *King James Bible.*

Daedalus

61: "everywhere I look, bodies emerge" – Private Jacques LaPointe. "In a flooded trench, the bloated bodies of some German soldiers are floating. Here and there, too, arms and legs of dead men stick out from the mud, and awful faces appear, blackened by days and weeks under the beating sun. I try to turn from these dreadful sights, but everywhere I look, bodies emerge…"

61: "no more unsuitable spot" – Colonel C.D. Baker-Carr, brigade commander, British Tank Corps: "To anyone familiar with the terrain in Flanders it was almost inconceivable that this part of the line should have been selected. If a careful search had been made from the English Channel to Switzerland, no more unsuitable spot could have been discovered."

61: "send us no more" – from member(s) of Sir Hubert Gough's staff, Fifth Army Headquarters (mentioned by Lloyd George in his memoirs).

62: "I hate the idea of thrusting an army" and Haig's response – Colonel Repington's diary for June 24, 1917.

62: "perhaps Haig, listening to his God" – on June 24th, 1916, the first day of the British bombardment of the Somme, Haig wrote: "I feel that every step in my plan has been taken with the Divine Help."

62: "It was dark, sky lit with flares" – the details in this paragraph as well as the quote, "There came, with time, a sort of deadening," are all from CBC interviews with Canadian soldiers who fought in Passchendaele. These were aired in 1964, and led by CBC broadcaster J. Frank Willis. They were later broadcast in episode 4 of "The Bugle and the Passing Bell" – *Ideas with Paul Kennedy*, June 25, 2015.

63: "as Gaia in that book" – *D'Aulaire's Book of Greek Myths.*

64: "oh darling mother…" "armed for battle" from Ovid, *Metamorphoses* (Book 6, line 626 and 674) – translated by David Raeburn.

64: "And these are they" – Leviticus 11:13-19, *King James Bible.* The original reads "lapwing" but most other translations instead list "the hoopoe."

65: "Armless, boneless" – Joseph B. Geoghegan, "Johnny I hardly knew ye" (first published 1867). The full verse reads:

Ye haven't an arm, ye haven't a leg, hurroo, hurroo
Ye haven't an arm, ye haven't a leg, hurroo, hurroo,
Ye haven't an arm, ye haven't a leg
Ye're an armless, boneless, chickenless egg

Ye'll have to put with a bowl out to beg
Oh Johnny I hardly knew ye.

65: "the summer that came before the war" – imagery of warm weather, picnics, white wicker table under trees, as well as Sassoon hunting foxes – from Paul Fussell's *The Great War and Modern Memory*. Geoff Dyer also mentions this summer, noting that "The glorious summer of 1914 seems, even, to have been generated by the cataclysm that succeeded it."

66–67: "sobbing, but absent-mindedly… perhaps a cow's skull… great boots… Old Jevons… sky sullen grey" – Virginia Woolf, *Jacob's Room.*

66–67: "The war tainted the past…" and quote of Wyndham Lewis' – Geoff Dyer, *The Missing of the Somme.*

67: "dead sea of mud… skulls appear" – Edmund Blunden, *Undertones of War.*

67: "We heard the guns" – Stuart Cloete.

68: "slew his son" – Wilfred Owen, "The Parable of the Old Man and the Young."

68: "The trenches are a labyrinth…" – Major Frank Isherwood, in a letter to his wife, December 1914.

68: "I used to hear the larks" – Sergeant Major F.H. Keeling.

68: "Good God… it's worse" – This was described by David Lloyd George in his memoirs. He did not mention names, but it has since been attributed to Lieutenant-General Sir Launcelot Kiggell, with the man who responded having fought in Passchendaele. This story has become famous over the years, but some doubt its veracity.

Medusa

69: "flame licks all over… deadly white… blaze of colour" – C.S. Lewis, *The Lion, the Witch and the Wardrobe.*

70: "as Faulkner said" – "The past is never dead. It's not even past." – William Faulkner, *Requiem for a Nun.*

71: "of all the teeming perils…" – Angela Carter, "The Company of Wolves" from *The Bloody Chamber*. "Never stray from the path… never eat…" from the screenplay (co-written with Neil Jordan, 1984).

71: "spiders, having parts extending" – *Maryland Gazette*, May 12, 1816.

71: "world will end July 18th" – an astronomer in Bologna, described as a "mad Italian prophet" by *The Times*, made this prediction in late spring.

71: "killed nothing, as all were dead before" – *Connecticut Courant,* October 1, 1816 – correspondent from Danville, South Carolina.

72: "though there are no clouds" – *Connecticut Courant,* October 15, 1816 – William Young, teacher in Plattsburg, NY.

72: "the wind blows cinders onto vessels" – The smoke from these fires supposedly caused several shipwrecks.

73: "standing corn" – Virgil. Shepherd Daphnis: "these herbs and poisons plucked in Pontus, Moeris himself gave me as a gift. Helped by these I have often seen Moeris change into a wolf, summon up the dead from the depths of the grave, and move the standing corn into other men's fields."

73: "had very strangely ruined" – Henri Boguet, "Of the Metamorphosis of Men into Beasts and Especially of Lycanthropes or Loups-garoux" in *Discours De Sorciers* (1590).

73: "by moving the inner perceptions…" – Heinrich Kramer (Henricus Institoris), "Question X. Whether Witches Can by some Glamour Change Men into Beasts" *Malleus Maleficarum* (first published in 1487) translated by Montague Summers.

74: "there is a photograph of a tree… Ferris wheel… mining town… Chernobyl" – series of pictures found on www.boredpanda.com/nature-reclaiming-civilization/. The piano (at California State University, Monterey) was actually a stage prop, cut apart then placed around the tree, by artist Jeff Mifflin (the installation is called "Piano Tree"). The tree carried on growing. The Ferris wheel (North Carolina) and mining town (in Namibia) appear to have occurred naturally—those photos are credited to Kyle Telechan and Marsel van Oosten, respectively. I could not find who took the Chernobyl shot—it is a view from above of the ghost town Pripyat in Ukraine.

74: "…it was a question of botany" – Henrich Boll, *The Silent Angel.* Boll does not explicitly write that the city is Cologne.

75: "ladies' dresses" – in Jonathan Miles' *The Wreck of the Medusa: The Most Famous Sea Disaster of the Nineteenth Century,* he writes that the men were "decked in finery." It may not have been ladies' dresses.

75: "at eleven come the wolves" – from Barry Lopez's *Of Wolves and Men.*

78, 84: "while elsewhere" – details of the ships and captain, voyage length, natural hazards, etc. from the *Trans-Atlantic Slave Ship Database*, www.slavevoyages.org.

80: "It was impossible" – Julian Barnes, "Shipwreck" (chapter 5), *A History of the World in 10½ Chapters.*

81: "hack your flesh away" – Homer, Book XXII, line 409, *The Iliad*, translated by Robert Fagles.

81: "I could be a God rising up in anger… at Achilles, who blocks the waterways with his dead." – Achilles' fight with Scamander – Homer, Book XXI, *The Iliad.*

82: "One, two. Why, then…" – William Shakespeare, *Macbeth*, Act 5, Scene 1.

The Children of Famine

85: "Look at that man on that hill… sailing above" – Ovid, *Metamorphoses* (Book 1, line 293–296), translated by David Raeburn.

85: "rotting grain, pigs' ears, bread made from sawdust and straw, eating pet dog" – details from contemporaries in southern German states.

86: "In Württemberg" – by the end of 1816, grain prices here had doubled from that of the year before; between November 1816–July 1817, the price of oats had tripled and potatoes quadrupled. "…cadavers trudged through the streets" – imagery from a contemporary, who saw "persons who looked like cadavers, and among them multitudes of children crying out for bread."

86: "Das Hungerjahr" – nickname from Switzerland.

86–88: "I've got to kill you… not enough to still their hunger… Now it's your turn… you've got to die or else [I'll] waste away (original was "we'll")… "Dear mother, we'll lie down and sleep…" as well as slightly altered "Oh, dear mother… spare me…" from The Brothers Grimm's "The Children of Famine" (translated by Jack Zipes). In this seven-paragraph-long story, the two daughters both bring back "a little piece of bread." In the end, the mother "departed, and nobody knows where she went." This tale was published in Volume II (1815). The source is Johannes Praetorius, *Der abentheuerliche Glückstopf* (1669).

86: "Finally, in Laichingen" – in this town, where almost 80% couldn't afford to buy bread or grain, the officials refused to distribute free wheat from the state granary until the townspeople threatened a violent hunger march. The rich bought up property and withheld donations to the poor relief fund.

87: "thousands of men and women, ripping chunks out of a living chestnut mare… They're sending our corn out to Switzerland" – Riots broke out in Augsburg, Memmingen and other towns in the winter of 1817, over the rumoured export of corn to Switzerland, while the townspeople ate horse and dog flesh. (*Times* January 1, 1817).

Let Not the Lord

92: "dream that my little baby" – Mary Godwin (later Shelley), in her journal (March 19, 1815).

93: "I wish that supposing I fly well… Since last night… I have now decided…" – letters (dated January 8, April 2, and April 8, 1828) from Augusta Ada King-Noel, Countess of Lovelace (better known now as "Ada Lovelace") to her mother, Anne Isabella Millbanke ("Isabella"). Byron and Isabella separated one month after Ada's birth, and she never saw her father again. Ada Lovelace went on to work in mathematics, and is now widely recognized as the first computer programmer. She died at the age of 36.

95: "This day was a verie great floud" – from clan chief and politician Alexander Brodie's diary. The full text of *The Diary of Alexander Brodie of Brodie…* can be found online at archive.org.

96: "Eli, Eli" – "And about the ninth hour Jesus cried with a loud voice, saying, *Eli, Eli, lama sabachthani?* that is to say, My God, my God, why hast thou forsaken me?" – Matthew 27:46, *King James Bible.*

96: "With slow but irresistible power" – Louis Simond.

96–97: "Never was a scene more awfully desolate" – Mary Shelley *History of a Six Weeks' Tour through a part of France, Switzerland, Germany and Holland (1817)*

98: "There are troops of wolves" – Percy Bysshe Shelley in a July 28th, 1816 letter to Thomas Love Peacock.

99–100: "threw down the chimney pots" and the rest of the description of the August 16, 1816 earthquake in this section – taken from contemporary accounts found on Scottish Archives for Schools website (http://www.scottisharchivesforschools.org/naturalScotland/Earthquake.asp). "Threw down the chimney pots" comes from Groome's Gazetter. All other quotes come from Sir Thomas D. Lauder's "Account of the Earthquake of 1819" and newspaper accounts, both of which were collected by David Milne and recorded in his *Notices of Earthquake-shocks felt in Great Britain, and especially in Scotland, with inferences suggested by these notices as to the causes of these shocks* (Edinburgh New Philosophical Journal, July 1841).

100: "The rain… poured" – Mary Shelley, *Frankenstein.*

100: "Lower part… was entirely inundated…" – Lady Caroline Capel, July 2016 letter.

100: "The great increase of beggars" – Sir Thomas Stamford Raffles, *Letters during a Tour through Some Parts of France, Savoy, Switzerland, Germany and the Netherlands, in the Summer of 1817.* "The only unpleasant circumstance in crossing the Jura, and which bespoke the deep poverty of the people, was the great increase of beggars. They were chiefly children, and their numbers and their impunity were truly astounding… we were glad when a little level road allowed us to go on at a quick rate, and thus lose, for a while, the distressing din."

100: "It is terrifying to see these walking skeletons…" – quoted in Wood.

101: "My dearest Hogg" – Mary Godwin to Thomas Jefferson Hogg, March 6, 1815.

101: "her third pregnancy" – Daisy Hay, *Young Romantics: The Shelleys, Byron and Other Tangled Lives.*

102: "the floods of rain" – *Vita Edwardi Secundi: The Life of Edward the Second*, written in Latin by an unknown English historian, probably in 1326.

103: "Fields full" – ergotism caused hallucinations, manic behaviour, fever, and/or gangrene. The ergot infection of grains such as rye happened more often in times of rain because the rye flower stayed open longer, thereby providing more time for the fungus to infect it.

103: "Poor gnaw like dogs…" – original quote is "gnawed, just like dogs, [on] the raw, dead bodies of cattle…" "grazed like cows…" – from Fritz Curschmann's *Hungersnöte im Mittelalter.*

The Ice Maiden

106: "There is a story of an artist" – Bartholomäus Traubeck. Some of the description in this paragraph (and the one beginning "and the tree ring was so faint") is adapted from Andrea Peltser's essay, "By Way of Beginning," in her collection *Limber.* "Even a tree" is directly quoted from this essay.

106: "The procession came here two hundred years ago." – In 1653, two Jesuit priests led a procession on a four-hour walk to a glacier. They sang psalms and hymns en route and, when they arrived, the Jesuits celebrated mass, preached a sermon, and then sprinkled the front with holy water. In 1664, the bishop of Geneva led a procession of 300 people to where another glacier, Des Bois, encroached on the village Les Bois. He blessed the glacier as well as a ring of ice sheets.

106: "Then this year of Our Lord" – By 1818, the Bossons glacier had submerged five hectares of farmland and threatened the village of Monquart. In response, villagers planted a large cross at its end.

106: "'Vanity,' she laughed" – The Ice Maiden doesn't actually say this, but it's implied: "Vermin is what you are! Just one tumbling snowball and you and your houses and towns would be crushed and obliterated!... There they sit, all those *ideas!* But they're caught in the grip of nature..." (Hans Christian Andersen, "The Ice Maiden," translated by Tiina Nunnally).

107: "the echo of church bells... they're pretending to be gods...." – Hans Christian Andersen, "The Ice Maiden."

107: "this earth which we inhabit" – Shelley in a July 22nd, 1816 letter to Thomas Love Peacock, referencing French naturalist Georges-Louis Leclerc, Comte de Buffon's *La Théorie de la Terre* (1749) which posited that the world would end in a new ice age: "I will not pursue Buffon's sublime but gloomy theory, that this earth which we inhabit will at some future period be changed into a mass of frost..."

107: "There is a dam made of ice" – the ice dam was created by the Giétro glacier in 1818. If it burst, it could drown the town of Martigny in the Val de Bagnes. In response, the government sent for engineer and glaciologist Ignace Venetz. Venetz ordered the cutting of the tunnel through the ice wall. The workers, Italian men, worked day and night, without proper clothing, including waterproof boots. Many quit. Venetz then hired Swiss montagnards who worked for another month as the water level continued to rise. On June 16th, the wall of ice crashed down and the river Dranse burst forth. Without this tunnel, the disaster would have been considerably worse.

108: "so beautiful and so populous" – H.C. Escher, "A Description of the Val de Bagnes in the Bas Valais, and of the Disaster which Befell It in June, 1818" *Blackwood's Edinburgh Magazine* (October 1818–March 1819).

108: "There is nothing, the Ice says" – Ursula LeGuin, *The Left Hand of Darkness.* "There is nothing, the Ice says, but Ice. But the young volcano there to northward has another word it thinks of saying."

The Laufmashine

109: "three-wheeler with a second seat," "sails for windy days," "connected to the rear axle," "steering by means of a tiller" as well as von Drais' being "disheartened" and going back to Baden to invent other things is from David V. Herlihy's *Bicycle: The History*. Critic who said the machine turned "a man into a horse and carriage" – *Baltimore Morning Chronicle,* May 19, 1819.

111: "because, says one writer, There were no horses left… we have no statement… thus, there was a need for a horseless means of transport" – actually two authors, Tony Adland and Hans-Erhard Lessing, in their *Bicycle Design: An Illustrated History.*

111: "with nothing alive but her and a jewel-eyed snake" – Frances Hodgson Burnett, *The Secret Garden:* "when she looked down she saw a little snake gliding along and watching her with eyes like jewels… He slipped under the door as she watched him.

'How queer and quiet it is,' she said. 'It sounds as if there was no one in the bungalow but me and the snake.'"

112: "this invention, says another…" – Soren Fink "with scientific support of Prof. Dr. H.E. Lessing" www.karldrais.de.

113: "the main idea of the invention" – this was Drais' only word on the matter.

113, 115: "I am in the garden reading a memoir" and the later quote, "in this brightly lit, alien world" – Vera Brittain, *Testament of Youth.*

114: "A U.S. study" – scientists David Luther and Elizabeth Derryberry compared the differences between modern birdsong of the male white-crowned sparrow and those of recordings in the same area from 1969. They also set up an iPod speaker in a park and played the songs recorded in 1969 and 2005. The birds responded more strongly to the 2005 songs.

114: "illumination with thousands of lamps… temple for the music…" – Herman Ludwig Pückler-Muskau (Prince of Muskau), 1827.

115: "only gull that regularly nests… open cup…" entry on Bonaparte's Gull, *Cornell Lab of Ornithology*, www.allaboutbirds.org

115: "the sound… disturbed at the nest…" – J. Fenwick Landsdowne's *Birds of the West Coast; Volume One.*

116: "the rain was falling in torrents" – *Napoleon's Memoirs.*

118: "white light of bursting shells… above all…" – General Seely, on the battle of the Somme.

118: "trying desperately to get to its feet" – diary of Lieutenant R G Dixon, 14th Battery, Royal Garrison Artillery.

118: "nearly every green thing… water running from their mouths… smoking entrails" along with much of the imagery of this and a few other sections comes from Ernst Junger's *Storm of Steel,* translated by Basil Creighton.

118–124: "It was all a ghost story… in trampled gardens… in the low hedges… the heavy machine… swollen with dead mules… with full-sized eels… the quality of the Somme mud… Let us stop this war" as well as other imagery – Edmund Blunden's *Undertones of War.*

118–119: "haunt burying-places… some that have been catched" – Titus Lewis's *Welsh-English Dictionary* (1805).

119: "a matter of holes and ditches" – Siegfried Sassoon *Memoirs of an Infantry Officer.*

119: "women rushed towards the cadaver" – Danish actress Asta Nielsen.

121: "Horses are slow and quickly" – "The Use of Three-Speed Bicycles in the War" *Bicycling World and Motorcycle Review,* January 19, 1915.

121: "…have dominion over the fish" – Genesis 1:26, *King James Bible.*

121: "many hundreds of dogs, carrier pigeons" – RSPCA memorial plaque.

121: "the horses… screaming" – Lieutenant Wheatley, *Officer and Temporary Gentleman.*

122: "almost perpendicular hills… great masses of quartz" – Peter Forbath's *The River Congo: The Discovery, Exploration and Exploitation of the World's Most Dramatic River.*

122: "skeletons dried up like mummies": Léopold Courouble, *En Plein Soleil: Les Maisons du Juge – Le Voyage à Bankana.* Courouble was a Congo state official. In his memoirs, he writes, "A file of poor devils, chained by the neck, carried my trunks and boxes toward the dock." At the next stop, more porters were needed: "There were about a hundred of them, trembling and fearful before the overseer, who strolled by whirling a whip. For each stocky and broad-backed fellow, how many were skeletons dried up like mummies…."

122: "When you fall off the bridge" – when "libérés chained by the neck cross a bridge, if one falls off, he pulls the whole file off and it disappears." – from Jules Marchal's *L'Etat Libre du Congo: Paradis Perdue. L'Histoire du Congo 1876–1900, vol. I.*

122: "they bring Captain Rom twenty-one heads" – Léon Rom came to the Congo in 1886, at the age of 25. He became a district commissioner, then judge. He trained black troops for the Force Publique. Rom was also an avid collector of butterfly specimens and brought back specimens to Europe on his visits. He was elected a member of the Entomological Society of Belgium. Rom is also thought to be the real-life model for Conrad's Mr. Kurtz. Like Kurtz's collection of shrunken heads, Rom too collected the heads of "rebels" for decoration. In E.J. Glave's article "Cruelty in the Congo Free State" – (*The Century Magazine*, September 1897) he writes of the aftermath of a rebellion: "Many women and children were taken, and twenty-one heads were brought to the falls, and have been used by Captain Rom as a decoration round a flower-bed in front of his house!" It is possible that Rom and Conrad met—but even if not, it's quite likely that Conrad would have read this account, either in *The Century Magazine*, or when it was reprinted in the December 17, 1898, issue of *The Saturday Review*, a magazine Conrad was known to read.

122: "I always have to cut off" – a chief to William Sheppard and quoted in *The Missionary* (February 1900).

122: "To gather rubber in the district" – Charles Lemaire.

123: "the country a desert"… – Roger Casement, diary entry for June 5, 1903.

123: "When I crossed the stream…" – Swedish missionary E.V. Sjöblom, speech in London, May 12th, 1897.

123: "An ibis" – "Hippo downstream. Saw three pelicans feeding, close to us. Also saw a beautiful Egyptian ibis, black body, white wings; a lovely fellow in full flight over us." – Casement diary entry July 2, 1903.

123: "René de Permetier" – officer in the Equator district in the 1890s. His nickname, by Africans, was Bajunu (*bas de genoux)* because he made people kneel before him. "If he found a leaf in a courtyard that women prisoners had swept, he ordered a dozen of them beheaded. If he found a path in the forest not well-maintained, he ordered a child killed in the nearest village."

123: Centralia – in Centralia, Pennsylvania, a coal mine fire has been burning under the borough since 1962.

124: "The sheer devastation of the battlefield" – Paul Johnson, *The Birth of the Modern World: Society 1815–1830.*

124: "The farm of Hougoumont" – John Wilson Croker.

Afterword: The Path

125: "behind the glass" – descriptions throughout of displays at the Royal Ontario Museum.

Acknowledgements

This is a book of correspondences and connections, and therefore the connections are many, probably too extensive to list here. It certainly could not have been written without the use of many writers who have come before. Many of these, though probably not all, are listed here or in the notes.

My thanks in particular to Hans Christian Andersen (*Fairytales*, translated by Tiina Nunnally), Frances Hodgson Burnett, Lewis Carroll, Angela Carter, Ingri and Edgar Parin D'Aulaire, The Brothers Grimm (*The Complete First Edition: The Original Folk & Fairy Tales of the Brothers Grimm* and *The Complete Fairytales of the Brothers Grimm*, both edited and translated by Jack Zipes), C.S. Lewis, and Ovid. My years of rereading myths, nursery rhymes, children's literature, and fairy tales meant that they seeped in when I didn't always intend it. My life is richer for this.

Along with the sources already listed in the notes, the following were especially helpful for providing information and quotes: Steven R. Bown's *Scurvy: How a Surgeon, a Mariner, and a Gentleman Solved the Greatest Medical Mystery of the Age of Sail,* Simon Butler's *The War Horses: The Tragic Fate of a Million Horses in the First World War*, Pauline Couture's *Ice: Beauty, Danger, History,* Peter Davidson's *The Idea of North,*

Kathryn A. Edwards, ed. *Werewolves Witches and Wandering Spirits: Traditional Belief & Folklore in Early Modern Europe,* Will Ellsworth-Jones' *We Will Not Fight: The Untold Story of World War One's Conscientious Objectors,* James Essenger's *Ada's Algorithm: How Lord Byron's Daughter Ada Lovelace Launched the Digital Age,* Brian Fagan's *The Little Ice Age: How Climate Made History 1300–1850*, Joel Greenberg's *A Feathered River Across the Sky: The Passenger Pigeon's Flight to Extinction,* Margaret Guroff's *The Mechanical Horse: How the Bicycle Reshaped American Life,* Adam Hochschild's *King Leopold's Ghost: A Story of Greed, Terror, and Heroism in Colonial Africa,* Richard Holmes's *Shelley: the Pursuit,* Kathleen Jamie's *Sightlines: A Conversation with the Natural World,* William K. Klingaman and Nicolas P. Klingaman's *The Year Without Summer: 1816 and the Volcano that Darkened the World and Changed History,* Norman Leach's *Passchendaele: Canada's Triumph and Tragedy on the Fields of Flanders,* Kate Luard's *Unknown Warriors: The Letters of Kate Luard, RRC and Bar, Nursing Sister in France, 1914–18,* Robert MacFarlane's *The Old Ways: A Journey on Foot,* Charlotte F. Otten (ed). *A Lycanthropy Reader: Werewolves in Western Culture,* John Pollard's *Wolves and Werewolves,* Angelo S. Rappoport's *The Sea, Myths and Legends,* William Rosen's *The Third Horseman: Climate Change and the Great Famine of the Fourteenth Century,* W.G. Sebald's *On the Natural History of Destruction,* Henry Stommel and Elizabeth Stommel's *Volcano Weather: The Story of 1816, the Year Without a Summer,* David Thomson's *Nairn in Darkness and Light,* Leon Wolff's *In Flanders Fields Passchendaele 1917,* Gavin Weightman's *London's Thames,* C.E. Wood's *Mud: a Military History,* and Gillen D'Arcy Wood's *Tambora: The Eruption that Changed the World.*

Thank you to Nicholas Hughes and Jeremy Banning (http://jeremybanning.co.uk) for the use of their aerial photographs of (respectively) Tambora and Lochnagar Crater (from the Battle of the Somme).

Thank you to all those who participated in my social media editing queries. Especial thanks to Wilson Bell and Amy McLaughlin, for their particularly astute suggestions.

Many of these essays first found homes in journal or chapbook publication: "Their Useless Wings" (Baseline Press, 2015), "Mud" (*The New Quarterly*, Summer 2014), "Tambora" (*The Fiddlehead,* Winter 2015 and republished in Tinderbox Editions' *Braided Rivers: An Anthology of Lyric Essays,* 2018), "The Last Frost Fair" *(The Malahat Review*, Winter 2014), "Daedalus" (*Canadian Notes & Queries*, Summer 2016), and "Let Not the Lord" (*Arc Poetry Magazine,* Winter 2016). Especial thanks to John Barton, Jenny Haysom, Karen Schindler, and Susan Scott, for their edits.

Thank you to the Canada Council for the Arts and Arts Nova Scotia, for financial support.

Thank you to Ginger Pharand, whose declaration that werewolves did not encapsulate true transformation helped inspire the beginning of "Medusa."

Thank you to my mother, Sharon Yandle, and my sister, Carlyn Yandle, whose ongoing support of my writing means a great deal.

Thank you to the team at Biblioasis, and especially to guest editor Stephanie Bolster, whose support and advice on essay order, word choice, and epigraph was invaluable.

"The Frost Fair" was written in part with Zach Wells in mind, and so is, in part, dedicated to him. I can't express how much your friendship and support and long long stretch of love has meant to me over the years, and especially in the writing of this. More prosaically, thank you for the books you bought for or recommended to me in hopes they might be useful (most were!)

This book is for Kaleb. There is no glass.

PHOTO: NANCY MCCARTHY

Rachel Lebowitz is the author of *Hannus* (Pedlar Press, 2006), which was shortlisted for the 2007 Roderick Haig-Brown Regional Prize (BC Book Prize) and the Edna Staebler Award for Creative Non-Fiction. She is also the author of *Cottonopolis* (Pedlar Press, 2013) and the co-author, with Zachariah Wells, of the children's picture book *Anything But Hank!* (Biblioasis, 2008, illustrated by Eric Orchard). She lives in Halifax, where she coordinates adult tutoring programs at her neighbourhood library.